ANNALS OF MATHEMATICS STUDIES

Number 25

ANNALS OF MATHEMATICS STUDIES

Edited by Emil Artin and Marston Morse

1. Algebraic Theory of Numbers, *by* HERMANN WEYL
3. Consistency of the Continuum Hypothesis, *by* KURT GÖDEL
6. The Calculi of Lambda-Conversion, *by* ALONZO CHURCH
7. Finite Dimensional Vector Spaces, *by* PAUL R. HALMOS
10. Topics in Topology, *by* SOLOMON LEFSCHETZ
11. Introduction to Nonlinear Mechanics, *by* N. KRYLOFF *and* N. BOGOLIUBOFF
14. Lectures on Differential Equations, *by* SOLOMON LEFSCHETZ
15. Topological Methods in the Theory of Functions of a Complex Variable, *by* MARSTON MORSE
16. Transcendental Numbers, *by* CARL LUDWIG SIEGEL
17. Problème Général de la Stabilité du Mouvement, *by* M. A. LIAPOUNOFF
19. Fourier Transforms, *by* S. BOCHNER *and* K. CHANDRASEKHARAN
20. Contributions to the Theory of Nonlinear Oscillations, Vol. I, *edited by* S. LEFSCHETZ
21. Functional Operators, Vol. I, *by* JOHN VON NEUMANN
22. Functional Operators, Vol. II, *by* JOHN VON NEUMANN
23. Existence Theorems in Partial Differential Equations, *by* DOROTHY L. BERNSTEIN
24. Contributions to the Theory of Games, Vol. I, *edited by* H. W. KUHN *and* A. W. TUCKER
25. Contributions to Fourier Analysis, *by* A. ZYGMUND, W. TRANSUE, M. MORSE, A. P. CALDERON, *and* S. BOCHNER
26. A Theory of Cross-Spaces, *by* ROBERT SCHATTEN
27. Isoperimetric Inequalities in Mathematical Physics, *by* G. POLYA *and* G. SZEGO
28. Contributions to the Theory of Games, Vol. II, *edited by* H. KUHN *and* A. W. TUCKER
29. Contributions to the Theory of Nonlinear Oscillations, Vol. II, *edited by* S. LEFSCHETZ
30. Contributions to the Theory of Riemann Surfaces, *edited by* L. AHLFORS *et al.*

Contributions to Fourier Analysis

A. ZYGMUND

W. TRANSUE

M. MORSE

A. P. CALDERON

S. BOCHNER

PRINCETON

PRINCETON UNIVERSITY PRESS

1950

London: Geoffrey Cumberlege, Oxford University Press

The work of preparing Contributions III, and IV in this volume was supported in part by The Office of Naval Research, and reproduction of these two Contributions in whole or in part for any purpose of the United States Government will be permitted.

Princeton University Press books are printed on acid-free paper and meet the guidelines for permanence and durability of the Committee on Produciton Guidelines for Book Longevity of the Council on Library Resources.

Printed in the United States of America

CONTENTS

CONTRIBUTIONS TO FOURIER ANALYSIS

I. LOCALIZATION OF BEST APPROXIMATION

By S. Bochner

Introduction. In the theory of convergence and summability whether for ordinary Fourier series or other expansions emphasis is placed on the phenomenon of localization whenever such occurs, and in the present paper a certain aspect of this phenomenon will be studied for the problem of best approximation as well.

If for a Fourier series

$$(1) \qquad f(x) \sim \sum_m a_m e^{imx}$$

we consider the sequence of ordinary partial sums $\{s_n(x)\}$ or of some generalized partial sums, then for two functions $f(x)$, $g(x)$ having the same values in the neighborhood of a point x_0 the corresponding sequences $\{s_n(x_0)\}$, $\{t_n(x_0)\}$ will converge or diverge simultaneously. This is a non-specific version of the localization theorem. A more specific version states that the two sequences are then equi-convergent; or, what is the same, that if we consider only one function, and if it is zero in a neighborhood of x_0, then its partial sums are also convergent at the point and also towards the value 0. Thus, in the specific version the localization theorem is not only a comparison theorem as such but actually a convergence criterion in its own right, although on the face of it, a somewhat rudimentary one. In our context, this specific version of the localization theorem will be envisaged, but with the slight although important generalization of allowing the function to assume in the neighborhood of the point x_0. some arbitrary constant value, zero or not. We will also consider functions having several such intervals of constancy simultaneously, with the constant value of the function differing from interval to interval, and our construction will be such as to give the optimal approximation in all these intervals simultaneously, so that in the end a theorem of rather more "global" than "local" implications will result, first appearances notwithstanding. We will not encumber our analysis with letting the function be piecewise analytic in certain intervals, instead of only piecewise constant, but such an extension could probably be obtained.

We will algorithmically associate with each periodic function (1) a sequence of exponential polynomials $\sigma_n(x)$ which approximate to it in the manner of Fejér polynomials in all cases, and which at a point of constancy approximate to it with the very small error $O(e^{-n\cdot\epsilon(n)})$, uniformly in the size of the neighborhood of constancy; and we will point out that a similar result also holds for approximation of non-periodic functions by ordinary polynomials. In the latter case such polynomials can be constructed from different approaches. But we will demonstrate that no matter how constructed the maximal coefficient of the n-th polynomial will tend to $+\infty$ with n; whereas in the periodic case the coefficients of our trigonometric sums will be bounded for all n, for every function.

It has been first recognized by Dunham Jackson[1] that the measure of approximation to a continuous periodic function (1) by its Fejér sums

$$\sigma_n^1(x) = \sum_{-n}^{n} \left(1 - \frac{|\nu|}{n}\right) a_\nu e^{i\nu x} \tag{2}$$

that is

$$\sigma_n^1(x) = \frac{1}{2\pi} \int_{-\pi}^{\pi} \frac{\left(\sin \frac{nt}{2}\right)^2}{n\left(\sin \frac{t}{2}\right)^2} f(x+t)dt, \tag{3}$$

is not as good as one might wish for. If we introduce the module of continuity

$$\omega(t) = \max_{|x_1 - x_2| \leq t} |f(x_1) - f(x_2)| \quad , \tag{4}$$

then for large n the difference $f(x) - \sigma_n^1(x)$ is not majorized by $\omega(\frac{1}{n})$ itself, as one would expect, but only by the larger quantity $\omega(\frac{1}{n})|\log \omega(\frac{1}{n})$ and this is roughly speaking due to the fact that the denominator $(\sin t)^2$ in the Fejér kernel is of the order t^2 for small t, which is better than the order t in the Dirichlet kernel itself, but yet not good enough to give optimal approximation for a continuous function in general. Jackson remedied

1. See his inaugural dissertation "Ueber die Genauigkeit der Annäherung stetiger Funktionen durch ganze rationale Funktionen gegebenen Grades und trigonometrische Summen gegebener Ordnung", Göttingen 1911. In subsequent papers and in particular in his Colloquium Lectures "The Theory of Approximation", New York, 1930, Jackson has elaborated an approach which is a composite of his original one and of the version due to de la Vallée Poussin, but we think that the version of de la Vallée Poussin has not been superseded by that.

this situation by introducing the better kernel

$$\frac{(\sin nt)^4}{n^3(\sin t)^4} \tag{5}$$

and in this way he could also attack the problem of best approximation for functions having continuous derivatives of any prescribed order.

However the analysis of Jackson in his first paper was strangely cumbersome and involved in that he laboriously obtained an approximation by ordinary polynomials first and from this derived a trigonometric approximation afterwards which in retrospection appears to be the very reverse of what, with his important method at least, is syllogistically correct. The correction was very soon introduced by de la Vallée Poussin[2] who developed the idea of Jackson into a neat methodical procedure on which the present paper will also be based.

He first of all rewrote the integral (3) into the alternate version

$$\sigma_n^1(x) = \frac{1}{2\pi} \int_{-\infty}^{\infty} \frac{(\sin n\frac{t}{2})^2}{n(\frac{t}{2})^2} f(x+t)dt \tag{6}$$

whose technical advantage it is that it now subsumes under the general scheme

$$n \int_{-\infty}^{\infty} K(n(x-t))\ f(t)dt \equiv \int_{-\infty}^{\infty} K(t)\ f(x + \frac{t}{n})dt$$

in which the Fourier analytical dependence on n is a very simple one. In the version (6) the periodicity of the function will no longer enter explicitly, and it will only reappear in this manner that the approximating functions will then be periodic as well. Otherwise it will also give a result for almost periodic functions or for non-periodic functions having Fourier transforms; and it will be immediately adaptable to the multi-variable case, in which such formulas extending over the entire space are of considerable importance.[3]

---

2. See his book, "Leçons sur l'approximation des fonctions d'une variable reelle (Collection Borel)", Paris 1919.

3. We are referring to their occurrence in our "Spherical Summability of Multiple Fourier Series". Transactions of the Amer. Math. Soc. 40 (1936), pp. 175-207.

Starting with (6), de la Vallée Poussin then introduced for any integer r the expressions

$$\sigma_n^r(x) = \frac{1}{M(r)} \int_{-\infty}^{\infty} \frac{(\sin \frac{nt}{2})^{2r}}{n^{2r-1} \, (\frac{t}{2})^{2r}} \; f(x+t)dt,$$

which are again exponential sums, but of the increased order rn, and, by forming certain linear combinations of such, he obtained exponential polynomials which for a function having r continuous derivatives will approximate with a precision of order n^{-r}. Now, in our case, due to the constancy of the functions in intervals, the second step of forming linear combinations will not be needed, but on the other hand, since a constant function has derivatives of all orders, a passage to the limit, $r \to \infty$, will be called for. We will let both r and n tend to infinity simultaneously, with n going faster than r and in a tempo that will be prescribed by the demand of an optimal result. This type of diagonalization out of a succession of summability processes may be new, although a certain disposition towards this manner of reasoning may be read into the work of I. J. Schoenberg.[4]

I. Two Theorems. We will now state two theorems.

THEOREM 1. Take a sequence of positive numbers $\varepsilon(n)$ which are converging to 0 as $n \to \infty$, no matter how slowly. This given, we will algorithmically assign to any periodic function

$$f(x) \sim \sum_m a_m e^{imx}$$

of class L a sequence of exponential polynomials $\{s_n(x)\} = \{s_n(f;x)\}$, the subscript n always denoting a bound for the exponents, such that the following two properties are satisfied simultaneously.

(i) The sequence has all the features of a Fejér sequence: it approximates to f(x) in L_1-norm,

$$\int_0^{2\pi} |f(x) - s_n(x)| \, dx \to 0, \qquad n \to \infty, \tag{7}$$

4. See his "Contributions to the Problem of Approximation of Equidistant Data by Analytic Functions", Quarterly of Applied Mathematics, 4 (1946), pp. 45-97 and 112-141.

and also in L_p-norm or C-norm if f(x) belongs to such a class, and $s_n(x) \rightarrow f(x)$ at points of continuity or simple discontinuity, and for a bounded function f(x) we have

$$\inf_x f(x) \leqq s_n(x) \leqq \sup_x f(x). \tag{8}$$

(ii) If in an interval

$$a < x < b \tag{9}$$

the function happens to be a constant,

$$f(x) = c, \tag{10}$$

then we have

$$s_n(x) - c = o(e^{-n\ \varepsilon(n)}), \tag{11}$$

uniformly in every closed subinterval

$$a + \delta \leqq x \leqq b - \delta\ . \tag{12}$$

This is our approximation theorem proper. The order term in (11) cannot be improved upon, as will be inferred from the following counter theorem.

THEOREM 2. If for a sequence of exponential polynomials $\{\sigma_n(x)\}$ we have

$$|\sigma_n(x)| \leqq 1, \tag{13}$$

or more generally

$$\frac{1}{2\pi} \int_0^{2\pi} |\sigma_n(x)|\, dx \leqq 1, \tag{14}$$

and if for three constants $\alpha > 0$, $\epsilon_0 > 0$, $C_1 > 0$ we have[5]

5. The symbol $C_1, C_2, C_3, \ldots$ will be positive numbers which do not depend on n or parameters occurring, but may depend on the function f(x) given.

(15) $$|\sigma_n(x)| \leq C_1 e^{-n\epsilon_0}$$

for x in

$$-\alpha \leq x \leq \alpha \quad ,$$

then the sequence $\sigma_n(x)$ converges to 0 uniformly in all of $-\pi \leq x \leq \pi$.

II. PROOF of theorem 1. It will be appropriate to replace the discrete subscript n by a continuous parameter R, $0 < R < \infty$. Thus $s_R(x)$ will be an exponential sum $\sum a_m e^{imx}$ in which $|m| \leq R$. Similarly we will be dealing with a positive quantity $\epsilon(R)$, which $\to 0$ as $R \to \infty$, and (11) will be replaced by

(17) $$s_R(x) - c = o(e^{-R\ \epsilon(R)}) .$$

However we will introduce another parameter, r, and this will be an integer, $r = 1,2,3, \ldots$, so that even when assumed or constructed to be a function of R,

(18) $$r = r(R) \ ;$$

it will still be understood that

(19) r(R) is an integer.

We now take the Fejér kernel

$$F(t) = \left(\frac{\sin \frac{t}{2}}{\frac{t}{2}} \right)^2 , \qquad -\infty < x < \infty ,$$

or any continuous function F(t) having the following properties:

(20) $$F(t) = F(-t), \qquad F(t) \geq 0$$

(21) $$F(t) \leq 4t^{-2} , \qquad 1 \leq t \leq \infty$$

(22) $$F(t) \geqq e^{-at}, \qquad 0 \leqq t \leqq \infty$$

for some positive $a = C_2$, and, what is decisive,

(23) $$\int_{-\infty}^{\infty} F(t)e^{i\alpha t}\, dt = 0 \qquad \text{for } |\alpha| \geqq 1 .$$

For any integer r we introduce the number

(24) $$M(r) = \int_{-\infty}^{\infty} F(t)^r\, dt ,$$

so that (22) will imply

(25) $$M(r) \geqq \int_0^1 e^{-atr}\, dt \geqq C_3\, r^{-1}.$$

Finally we introduce the function

(26) $$F_r(t) = \frac{F(t)^r}{M(r)}$$

and its Fourier transform

(27) $$\varphi^r(\alpha) = \int_{-\infty}^{\infty} F_r(t)e^{i\alpha t} dt .$$

The constant (24) was introduced so as to make

(28) $$\int_{-\infty}^{\infty} F_r(t)dt = 1,$$

but the decisive property is

(29) $$\varphi^r(\alpha) = 0 \qquad \text{for } |\alpha| \geqq r,$$

and this one, by the rule of convolution,[6] is a consequence of (18); the

---

6. See S. Bochner and K. Chandrasekharan, "Fourier Transforms", Annals of Mathematics Studies, No. (9), Princeton, 1949, p. 5 and p. 58.

rule being that if

$$\varphi(\alpha) = \int_{-\infty}^{\infty} F(t)e^{i\alpha t}\,dt$$

$$\psi(\alpha) = \int_{-\infty}^{\infty} G(t)e^{i\alpha t}\,dt$$

then

$$\chi(\alpha) = \int_{-\infty}^{\infty} F(t)\,G(t)e^{i\alpha t}\,dt$$

has the value

$$\frac{1}{2\pi}\int_{-\infty}^{\infty} \varphi(\beta)\,\psi(\alpha-\beta)\,d\beta .$$

Now, with any periodic function (1) we associate the approximating functions

$$s_R^r(x) = \int_{-\infty}^{\infty} f(x+t)\,\frac{R}{r}\,F_r\,(\frac{R}{r}\,t)dt \tag{30}$$

for $0 < R < \infty$ and $r = 1,2,3, \ldots$, and it is easily seen that

$$s_R^r(x) = \sum_m \varphi^r(\frac{rm}{R})\,a_m\,e^{imx} , \tag{31}$$

so that (30) is an exponential polynomial of degree $\leq R$, no matter what r is. We now introduce a variable integer (18) and denote the resulting family

$$\left\{ s_R^{\,r(R)}(x) \right\} \tag{32}$$

simply by $\{s_R(x)\}$, and we are claiming the following

LEMMA 1. In order that, for $R \to \infty$, the family (32) have the property specified in part (i) of theorem 1, it is sufficient that

$$r(R) \to \infty \tag{33}$$

and

(34) $$r(R).R^{-1} \to 0$$

simultaneously, as $R \to \infty$.

PROOF. Due to normalization (28) the lemma will hold if $r = r(R)$ is such that for every fixed $\delta > 0$, the number

(35) $$\sup_{2\delta \leqq t < \infty} \frac{R}{r} F_r\left(\frac{R}{r} t\right),$$

which is a function of R, will tend to 0 as $R \to \infty$. Now, due to (34), for any given $\delta > 0$, for $R \geqq R_1$, we have $R^{-1} r \delta \geqq 1$, and hence for $t \geqq 2\delta$ we have

(36) $$\frac{R}{r} F_r\left(\frac{R}{r} t\right) \leqq C_4 \frac{R}{r}\left(\frac{R}{r}\delta\right)^{-2r} M(r)^{-1} \leqq C_5 R\left(\frac{R}{r}\delta\right)^{-2r},$$

by (21) and (25). Now, (33) and (34) imply

(37) $$R = r\,\psi(r)$$

where $\psi(r) \to \infty$ as $r \to \infty$, and hence, for sufficiently large R, we have

(38) $$\log R\left(\frac{R}{r}\delta\right)^{-2r} = \log R - 2r \log(\psi(r)\delta)$$

and for fixed δ and <u>r</u> sufficiently large this is

$$\leqq \log R - r \log \psi(r) \leqq \log r + \log \psi(r) - r \log \psi(r).$$

But the last expression $\to -\infty$ as $r \to \infty$; which proves the lemma.

LEMMA 2. Part (ii) of theorem 1 will be fulfilled if we define $\rho(R)$ by

(39) $$\frac{\log \rho(R)}{\rho(R)} = 2 \max\left\{ \epsilon(R), \frac{\log R}{R^{1/2}} \right\}$$

and then put

(40) $$r(R) = \frac{R}{\rho(R)} + \eta(R), \quad 0 \leqq \eta(R) < 1,$$

so as to make r(R) an integer.

PROOF. As for the definition of $\rho(R)$ we note that in the half-line $1 \leq \xi < \infty$ the function

$$\frac{\log \xi}{\xi} \tag{41}$$

is monotonely decreasing to 0 as $\xi \to \infty$. Therefore, for large R, there is a (unique) $\rho(R)$ satisfying (39) and

$$\rho(R) \to \infty \quad \text{as} \quad R \to \infty . \tag{42}$$

Also, since

$$\frac{\log \rho(R)}{\rho(R)} \geq 2\,\frac{\log R}{R^{1/2}}$$

and

$$\frac{\log R^{1/2}}{R^{1/2}} \leq 2\,\frac{\log R}{R^{1/2}} \quad ,$$

the monotoneity of (41) implies $\rho(R) \leq R^{1/2}$. Hence, for r(R) as defined by (40) we first of all have (33), and on the other hand we have

$$\varphi(R) \equiv \frac{R}{r(R)} \sim \rho(R) \tag{43}$$

Now, if $f(x) = c$ in an interval $x_0 - 2\delta \leq x \leq x_0 + 2\delta$, then (30) and (28) imply

$$| s_R(x_0) - c | \leq \int_{2\delta}^{\infty} | f(x_0 + t) + f(x_0 - t) - 2c | \frac{R}{r} F_r(\frac{R}{r} t)\, dt$$

and, by (43) and (21) this is

$$\leq C_6 R \sum_{n=0}^{\infty} (\frac{R}{r}(\delta + 2\pi n))^{-2r}$$

$$\leq C_7 \; R(\tfrac{R}{r}\;\delta)^{-2r} \; .$$

Now, for fixed $\delta > 0$ and large R, (43) implies

$$\frac{\log \varphi(R)\,\delta}{\varphi(R)} \sim \frac{\log \rho(R)}{\rho(R)}$$

and hence we have for large R,

$$-\log\,[R(\tfrac{R}{r}\,\delta)^{-2r}] = 2r \log\,(\tfrac{R}{r}\,\delta) - \log R = \frac{2R}{\varphi(R)} \log\,(\varphi\,(r)\,\delta) - \log R$$

$$\geq \frac{3}{2}\,\frac{R \log \rho(R)}{\rho(R)} - \log R \geq \frac{3}{2}\,\frac{R \log \rho(R)}{\rho(R)} - 2R^{1/2} \log R$$

$$\geq \frac{1}{2}\,\frac{R \log \rho(R)}{\rho(R)} + R[\frac{\log \rho(R)}{\rho(R)} - 2\,\frac{\log R}{\log^{1/2}}\,]$$

$$\geq R \quad \epsilon(R),$$

the last step due to (39). Altogether we obtain

(44) $$|\; s_R(x_0) - c \;| \leq C_8 \; e^{-R \;\; \epsilon(R)} \; .$$

Now, if $f(x) = c$ in (4), then C_8 may be chosen uniformly for x_0 in (12), and this completes the proof of lemma 2.

But the quantity (40) also validates lemma 1, and thus the proof of theorem 1 is completed.

III. PROOF of Theorem 2. We write explicitly

(45) $$\sigma_n(x) = \sum_{\nu=0}^{2n} a_{n\nu} \;\; e^{i(\nu-n)x}$$

and (14) implies

(46) $$|\, a_{n\nu} \,| \leq 1 \; .$$

If we now introduce the complex variable

$$z = e^t e^{ix} \tag{47}$$

and the ordinary polynomial

$$F_{2n}(z) = \sum_{\nu=0}^{2n} a_{n\nu} z^{\nu} ,$$

then for $|z| = 1$, that is $t = 0$, we have

$$| \sigma_n(x) | = | F_{2n}(z) | = | F_{2n}(z) z^{\rho} | \tag{48}$$

for any real constant ρ. Now, for fixed n, we put $\rho = -2n$, thus introducing the function

$$G_n(z) = \sum_{\nu=0}^{2n} a_{n\nu} z^{\nu-2n} = F_{2n}(z) z^{-2n} , \tag{49}$$

and we then consider the average

$$N_n(t) = \frac{1}{2\pi} \int_0^{2\pi} \log | G_n(e^t e^{ix}) | \, dx . \tag{50}$$

We claim that it is a convex function in $-\infty < t < \infty$, so that we have

$$N_n(t) \leqq \frac{t_2 - t}{t_2 - t_1} N_n(t_1) + \frac{t - t_1}{t_2 - t_1} N_n(t_2) \tag{51}$$

for

$$t_1 \leqq t \leqq t_2 . \tag{52}$$

In fact it is easily seen that (50) is a continuous function of t and that it suffices to prove (51) under the assumption that $\log |G(z)| \neq 0$ for

(53) $$\log |z| = t_1 \quad \text{or} \quad \log |z| = t_2 \quad .$$

Now, under the latter assumption we may put $G_n(z) = P(z)\, Q(z)$, where $P(z) \neq 0$ in (52) and $|Q(z)| \leqq 1$ in (52) and $|Q(z)| = 1$ for (53). It then follows that (51) holds for the average of $\log |Q(z)|$, and it also holds for $\log |P(z)|$ since the latter function is regular and harmonic in (52), and for such a function the average in an annulus is a constant; which proves (51).

In particular for $0 < t < T$ we have

(54) $$N_n(t) \leqq (1 - \frac{t}{T})\, N_n(0) + \frac{t}{T}\, N_n(T) \quad ,$$

and for fixed $t > 0$ we will let $T \to \infty$. Our function $G_n(z)$ is analytic at infinity and it easily follows that for large T the value $N_n(T)$ is bounded by some finite upper bound. Therefore, (54) implies the relation

(55) $$N_n(t) \leqq N_n(0)$$

which will be decisive.

The assumptions of theorem 2 imply

(56) $$N_n(0) \leqq -n \frac{\alpha \epsilon_0}{2\pi} + C_9$$

so that

$$N_n(t) \leqq - n \frac{\alpha \epsilon_0}{2\pi} + C_9,$$

and on recalling (49) and (50) we hence obtain

$$\frac{1}{2\pi} \int_0^{2\pi} \log |F_{2n}(e^t e^{ix})|\, dx \leqq - n \left\{ \frac{\alpha \epsilon_0}{2\pi} - 2t \right\} + C_9 \ ,$$

for $t > 0$. We now choose one fixed $t = t_0 > 0$, so that

$$\gamma = \frac{\alpha \epsilon_0}{2\pi} - 2t_0 > 0 \ ,$$

and we obtain

$$\frac{1}{2\pi} \int_0^{2\pi} \log |F_{2n}(e^{t_0+ix})| \, dx \leq - n\gamma + C_9 .$$

If we now apply the Poisson-Jensen inequality to the function $\log |F_{2n}(z)|$ in the circle $|z| \leq e^{t_0}$ we hence obtain

$$\log |F_{2n}(e^{ix})| \leq \frac{e^{t_0} - 1}{e^{t_0} + 1} \quad | - n\gamma + C_9 | \quad ,$$

and hence by (48) we have

$$| \sigma_n(x)| \leq C_{10} \, e^{-n\gamma_0}$$

for some $\gamma_0 > 0$. Which proves theorem 2.

IV. Rémarks.

1. Instead of a periodic function (1) we may take an almost periodic function of Bohr's type with an arbitrary Fourier expansion

$$(57) \qquad f(x) \sim \sum_\lambda a_\lambda e^{i\lambda x},$$

and then again consider the approximating functions (30). Each of the latter is again an almost periodic function,[7] and its expansion is

$$(58) \qquad \sum_\lambda \varphi^r\left(\frac{r\lambda}{R}\right) a_\lambda \; e^{i\lambda x} \quad ,$$

so that for the exponents λ actually occurring in (58) we have $|\lambda| \leq R$. We may also take as $f(x)$ a function of the Stepanoff class[7] S_p which is the narrowest generalization of the L_p-class from the periodic to the almost periodic case. For such a function we have, among others,

7. See S. Bochner, "Properties of Fourier Series of Almost Periodic Functions", Proc. Lond. Math. Soc. 26 (1927), p. 437.

(59) $$\left(\int_{\xi}^{\xi + 2\pi} |f(t)|^{p} \, dt \right)^{1/p} \leq C_{11}$$

and the approximating functions (30) are again Bohr almost periodic. Now it can be easily verified that for these functions the assertions and proof of theorem 1 are as before.[8]

2. Similarly we may consider in $-\infty < x < \infty$ a (non-periodic) function $f(x)$ for which a Fourier integral exists, say a function of class $L_1(-\infty, \infty)$. If we introduce the Fourier transform

$$\Gamma(\alpha) = \frac{1}{2\pi} \int_{-\infty}^{\infty} f(x) e^{-i\alpha x} \, dx$$

and defines $s_R^r(x)$ by (30), then we have

$$s_R^r(x) = \int_{-R}^{R} \varphi^r\left(\frac{r\alpha}{R}\right) \Gamma(\alpha) e^{i\alpha x} \, d\alpha$$

and an analogue to theorem 1 can be obtained. It also remains in force for a function having a generalized Fourier integral

$$f(x) \sim \int_{-\infty}^{\infty} e^{ix\alpha} \frac{d^k \Gamma_k(\alpha)}{d\alpha^{k-1}}$$

for any integer k according to our general definition.[9]

3. We may also consider functions in several variables, whether periodic, or almost periodic, or non-periodic. In Euclidean E_k we consider the function

(60) $$G(t_1, \ldots, t_k) = \Gamma(k) \, 2^k \, \pi^{-k} \, |t|^{-k} \, J_k(\pi t)$$

8. Perhaps in this connection we may recall, from the paper quoted in 7, the general "principle" there stated to the effect that any theorem known for periodic Fourier expansion has some kind of generalization to almost periodic expansions in which the arithmetical equi-distancing of the exponents is no longer stipulated; and that, whenever a theorem on periodic function is being based on an integral of type (3), the corresponding generalization will result from re-writing it into an integral of the type (6) first, and then obtaining the conclusion.

9. Bochner, "Vorlesungen ueber Fouriersche Integrale", Leipzig 1932, Chelsea 1948.

where $J_\alpha(x)$ is the classical Bessel function and where we put

$$|t| = (t_1^2 + \dots + t_k^2)^{1/2}$$

The Fourier transform

$$(61) \quad \varphi(\alpha_1,\dots,\alpha_k) = \int_{E_k} G(t_1,\dots,t_k)\, e^{i(\alpha_1 t_1+\dots+\alpha_k t_k)} dt_1 \dots dt_k$$

has, except for a constant factor, the value $(1 - |\alpha|^2)^{k/2}$ for $|\alpha| < 1$ and the value 0 for $|\alpha| \geq 1$. Now, the law of convolution also holds for functions in several vaiables and thus for the function $G(t)^2$ the Fourier transform vanishes for $|\alpha| \geq 2$. The function $F(t) = G(\frac{t}{2})^2$ has then the following properties:

$$(62) \qquad F(t) \geq 0$$

$$(63) \qquad F(t) \geq \exp(-a\,|t|), \quad \text{for } |t| \leq 1$$

$$(64) \qquad F(t) \leq (C_{12}^{-1}\, t)^{-k-1}, \quad \text{for } |t| \geq 1$$

and the presence of the constant C_{12} in (64), instead of 2 in (21) has the effect of replacing in the estimates the length 2δ by $C_{12}\delta$, whenever $C_{12} > 2$; and this would increase the constants occurring in our estimates whenever, for purpose of practical application, the explicit knowledge of such constants would be of interest. But otherwise the previous reasoning goes over in almost literal analogy. For any integers $r = 1,2, \dots$ we form the constants

$$M(r) = \int_{E_k} F(t)^r\, dt_1 \dots dt_k$$

and then the functions

$$F_r(t) = \frac{F(t)^r}{M(r)}$$

and their transform

$$\varphi^r(\alpha_1,\ldots,\alpha_k) = \int_{E_k} F_r(t)\, e^{i(\alpha_1 t_1 + \cdots + \alpha_k t_k)}\, dt_1 \ldots dt_k$$

and by the law of convolution we have again

$$\varphi^r(\alpha) = 0 \quad \text{for} \quad |\alpha| > r .$$

Now, if for any periodic function

$$f(x_1,\ldots,x_k) \sim \sum a_{m_1 \ldots m_k}\, e^{i(m_1 x_1 + \cdots + m_k x_k)}$$

we set up the functions

$$s_R^r(x) = \int_{E_k} f(x+t)\, (\tfrac{R}{r})^k\, F_r\, (\tfrac{R}{r}\, t)\, dt_1 \ldots dt_k$$

then they have the value

$$\sum_m \varphi^r(\tfrac{rm}{R})\, a_{m_1 \ldots m_k}\, e^{i(m_1 x_1 + \cdots + m_k x_k)}$$

and thus contain only such terms for which

$$(m_1^2 + \cdots + m_k^2)^{1/2} \leq R .$$

By judiciously choosing integers $r = r(R)$ we again obtain a sequence of exponential polynomials $s_R(x)$ having the following properties as $R \to \infty$. It approximates to $f(x)$ in an over-all fashion after the manner of Féjer polynomials, and if in an open set D the function has a constant value, $f(x) = c$, then in every compact subset D^o of D we again have the approximation (12); the choice of the polynomials depending only on $\epsilon(R)$ and nothing else.

4. We are returning to the one-dimensional case for a remark of a different trend. If $f(x) = c$ in (4) we may want to choose $r = r(R)$ so

as to make the difference $s_R(x) - c$ as small as possible in a variable interior

$$a + 2\delta(R) \leqq x \leqq b - 2\delta(R) ,$$

with $\delta(R) \to 0$, so as to sweep out the entire interior of (9) simultaneously with securing a good approximation. In this set-up we have to minimize the order of magnitude of

$$R\left(\frac{R}{r(R)} \ \delta(R)\right)^{-2r(R)} . \tag{65}$$

Now, if this is to tend to 0 as $R \to \infty$, we must have

$$\frac{R \ \delta(R)}{r(R)} \to \infty$$

and hence

$$R \, \delta(R) \to \infty ,$$

or $(\delta(R))^{-1} = o(R)$. Thus, $\delta(R)$ must not tend to 0 <u>too</u> rapidly. Suppose now we demand $\delta(R) = R^{-\alpha}$, $0 < \alpha < 1$, and we put tentatively $r(R) = R^{\beta}$, $0 < \beta < 1$. Then the logarithm of (65) is

$$\log R - 2(1 - \alpha - \epsilon)R^{\beta} \log R ,$$

and thus we see, that by a suitable choice of $r(R)$ we may arrange to obtain

$$s_R(x) - c = o(\exp(-R^{1-\alpha-\epsilon} \log R)) \tag{66}$$

in

$$a + R^{-\alpha} \leqq x \leqq b - R^{-\alpha} , \ 0 < \alpha < 1. \tag{67}$$

This remark also applies to the multi-dimensional case. If we take any open set of constancy then the estimate (66) applies to the variable subset whose distance from the boundary of the open set is $R^{-\alpha}$.

5. In this connection it should be noted that the last estimate and all previous estimates are valid for several intervals of constancy (or, in the multi-dimensional case, for several open sets of constancy) simultaneously. Thus, assuming, for instance, that our function is a step function throughout, that is piecewise constant in several intervals which together cover the interval of periodicity, then our construction gives a sequence of polynomials for which (8) holds, and for which at every point without exception there is convergence without Gibbs' phenomenon anywhere, and for which we have the estimate (17) in every fixed closed interior of an interval of constancy, or, if we prefer, the somewhat weaker estimate (66) in the sweeping-out interiors (67).

Also, everything also holds for almost periodic functions in $-\infty < x < \infty$ instead of only periodic ones, and for these we wish to emphasize the fact that the constants appearing then in our estimates may be chosen uniformly in the entire infinite line $(-\infty, \infty)$, and not only in every finite portion of it.

6. For computational purposes our exponential polynomials have an important structural "stability" which is also of some theoretical interest. If a function $G(t)$ in $-\infty < t < \infty$ is non-negative and

$$\int_{-\infty}^{\infty} G(t)dt = 1$$

then for its Fourier transform $\psi(\alpha)$ we have $|\psi(\alpha)| \leq 1$. If now for our previous choice of $r = r(R)$ we denote the quantity $\varphi^r(\frac{rm}{R})$ by $\lambda_{R,m}$ for $|rm| \leq R$, then for any function $f(x)$ with the expansion (1) we have

$$s_R(x) = \sum_{|m| \leq R} \lambda_{R,m}\, a_m\, e^{imx}$$

where

$$|\lambda_{R,m}| \leq 1 ,$$

so that $s_R(x)$ arises from $f(x)$ by insertion of uniformly bounded "multipliers"; and any upper bound for the Fourier coefficients of $f(x)$ will also be an upper bound for the Fourier coefficients of its approximations. This is in sharp contrast to what is liable to happen for approximation by <u>ordinary</u> polynomials as we are going to explain next.

7. Let $f(x)$ be given in $-1 \leqq x \leqq 1$. On putting $x = \cos \varphi$ we obtain a function $g(\varphi) = f(\cos \varphi)$, and intervals of constancy for $f(x)$ appear as such for $g(\varphi)$. If now we denote our previous approximating exponential polynomials of $g(\varphi)$ by $\sigma_n(\varphi)$, then the latter are linear combinations of $\cos \nu\varphi$, $\nu = 0,1,\ldots,n$. On putting

$$2 \cos \nu\varphi = \sum_\mu (-1)^{2\nu-\mu} \binom{2\nu}{\mu} (2 \cos \varphi)^{2\nu-\mu} ,$$

$\sigma_n(\varphi)$ goes over into an ordinary polynomial $s_n(x)$ of degree n in x, and the approximating properties which the sequence $\{s_n(x)\}$ inherits from the sequence $\{\sigma_n(\varphi)\}$ sound very much alike to those stated in theorem 1. However the coefficients of the sequence $\{s_n(x)\}$ are obviously no longer bounded in toto, and we are going to demonstrate that this is not an accident of the construction but must be inevitably so, except again in the trivial case in which the function $f(x)$ is a constant.

THEOREM 3. If the order functions $\rho(n)$, $\sigma(n)$ tend to $+\infty$ with n, and $\sigma(n)$ grows more slowly than $\rho(n)$, that is

$$\frac{\sigma(n)}{\rho(n)} \to 0, \qquad n \to \infty ;$$

if a sequence of analytic functions $s_n(x)$ of the complex variable x are analytic in the unit circle $|x| \leqq 1$ and $s_n(x) = O(e^{\sigma(n)})$ uniformly on the boundary $|x| = 1$; and if uniformly on a closed segment

(68) $$\alpha \leqq x \leqq \beta$$

of the open interval $-1 < x < 1$ we have $s_n(x) = O(e^{-\rho(n)})$, then the sequence $s_n(x)$ converges to 0 everywhere in $|x| < 1$, uniformly in every compact subset.

PROOF. We consider the doubly-connected domain bounded by $|x| = 1$ and (68) and we map it by a function $x = x(w)$ into an annulus

$$e^{t_0} \leqq |w| \leqq 1 ,$$

with the unit circumference mapping into itself. To the functions $f_n(w) = s_n(x(w))$ we apply the Poisson-Jensen inequality, and for $t_0 < t < 0$ we obtain

$$f_n(e^t e^{i\tau}) = O\{\exp(\sigma(n) - \frac{t}{t_o}\rho(n))\},$$

and our conclusions follow from the assumptions on the order functions.

If therefore a sequence of approximating polynomials

$$s_n(x) = a_{no} + a_{n1}x + \dots + a_{nn}x^n$$

is $O(e^{-n\epsilon(n)})$ on (68), then the sum

$$|a_{no}| + |a_{n1}| + \dots + |a_{nn}|$$

may not grow to infinity more slowly than $e^{n\epsilon(n)}$, -unless $f(x)$ is constant throughout -, which is a very much faster growth than boundedness of the coefficients a_m might bring about. On the other hand if the coefficients happen to be bounded, and $f(x)$ is not a constant everywhere then we cannot have $s_n(x) = O(n^{-\tau(n)})$ on (68) for any function $\tau(n)$ growing to $+\infty$, no matter how slowly.

II. DIRICHLET PROBLEM FOR DOMAINS BOUNDED BY SPHERES

By S. Bochner

Introduction. The so-called Dirichlet Principle is a means towards solving boundary value problems and is not really detachable from such, but it is also possible to give some precursory version of it in which the connection with boundary values has not yet been established. One such version is as follows.

If in an arbitrary domain S of the Euclidean E_k a given function $F(x_1, \ldots, x_k)$ of differentiability class C^1 has a finite Dirichlet norm

$$(1) \qquad \| F \| = \left(\int_S \left[\left(\frac{\partial F}{\partial x_1} \right)^2 + \ldots + \left(\frac{\partial F}{\partial x_k} \right)^2 \right] dv_x \right)^{1/2}$$

and if we consider the "distance"

$$(2) \qquad \| F - H \|$$

for all harmonic functions H having likewise finite Dirichlet norm, then this distance attains its absolute minimum, and the minimizing function H^0 is unique (up to a constant, of course). Also, for any other harmonic function H of finite norm we have

$$(3) \qquad (F - H^0, H) = 0$$

where

$$(4) \qquad (F,G) = \int_S \left(\frac{\partial F}{\partial x_1} \frac{\partial G}{\partial x_1} + \ldots + \frac{\partial F}{\partial x_k} \frac{\partial G}{\partial x_k} \right) dv_x ,$$

and this property also characterizes H^0; and finally we have the inequality

$$(5) \qquad \| F \| \geq \| H^0 \| .$$

For the sake of completeness we will give a succinct proof of this proposition in section 6 where we will also discuss the possibility of extending this proposition and theorem 1 from functions of class C^1 to classes D^1 consisting of "piecewise differentiable" functions as well; and we will briefly discuss the "Principle" for Banach norms other than (1).

The connection with boundary values is as follows. If the boundary B of the domain S is "sufficiently smooth", then for any continuous boundary function $f(\xi)$ on B there exists a harmonic function $H^1(x)$ assuming those values on its boundary. Now, $H^1(x)$ need not have finite Dirichlet norm, not even if S is the interior of a sphere, and on the other hand a function of finite norm need not have continuous boundary values, not even if it is harmonic. But if there does exist a function F(x) of finite norm having the continuous boundary values $f(\xi)$, then the function $H^1(x)$ just introduced must be the same as the minimizing function H^0 previously introduced, so that the solution H^1 of the boundary problem may then be obtained either by minimizing (1) or by solving (3), each time without reference to boundary values, or finally by minimizing $||F||$ for functions having the same boundary values.

The purpose of the present paper is to make this connection a more intimate one by attempting to eliminate the requirement that a <u>continuous</u> boundary function shall be prescribed. It seems to us that mere finiteness of the norm (1), or perhaps of some other similar norm, in a domain with a smooth boundary, automatically generates a certain type of boundary function which although not continuous is sufficiently tangible to allow the existence of a harmonic function with the same boundary function and to make this harmonic function unique. In the present paper we will not treat this problem in general but only for the Dirichlet norm in the very special case of a domain whose boundary consists of a finite number of spheres; and we wish to state that we place as much emphasis on theorem 2, or rather on its proof, as on theorem 1 itself.

The reader will note that for k = 2 conformal mapping will subsume a rather general class of domains under our seemingly very particular type. But for $k \geq 3$ no such device is available, and an extension to non-spherical domains would require a more searching kind of analysis.

Finally in section 7 we will also extend Douglas' solution of the problem of Plateau to boundary functions of our general description.

1. The Theorem.

THEOREM 1. <u>Assumptions</u>. Let S be a bounded domain in Euclidean E_k which is the interior of a sphere B, or, more generally, is bounded by a finite number of (k-1)-dimensional spheres $B_0, B_1, \ldots, B_p$, and let $F(x_1, \ldots, x_k)$ in S be a function of differentiability class C^1, for which the Dirichlet norm (1) is finite.

Conclusions. (i) There exists a boundary function of integrability class L_2 on every B_q, so that if, for instance, B_q is the sphere

(6) $$x_1^2 + \dots x_k^2 = 1$$

and if in polar coordinates

(7) $$r;\ \theta_1, \dots, \theta_{k-1}$$

our function F is an expression

(8) $$f(r;\ \theta_1, \dots, \theta_{k-1}),$$

then there exist almost everywhere on (6) a function

(9) $$f(1;\ \theta_1, \dots, \theta_{k-1})$$

so that

(10) $$\lim_{r \to 1} \int_{B_q} |f(r;\theta_1,\dots,\theta_{k-1}) - f(1;\ \theta_1,\dots,\theta_{k-1})|^2 \, dv_\theta = 0$$

holds.

(ii) There exists in S a harmonic function $H^0(x_1,\dots,x_k)$ of finite norm $||H||$ having the same boundary functions as F(x).

(iii) We have $||F|| \geq ||H^0||$, equality holding only for

(11) $$F = H^0 .$$

(iv) For every other harmonic function H of finite norm we have relation (3) and H^0 is the minimum of $||F - H||$ for all H of finite norm.

We note that the boundary functions (9) are functions of class L_2 relative to Euclidean (k-1)-dimension measure on Euclidean spheres, and that they constitute a proper subclass of such functions of class L_2 and not the entire class. We also note that the mere existence of such boundary functions could be obtained quite easily and that our more elaborate analysis is needed for showing that every element of the subclass is the boundary of a harmonic function as well.

Also, part (iii) of the theorem need not be proven independently since, once parts (i) and (ii) have been established, it is a consequence of (iv). But we have stated it separately because for the ordinary sphere it will follow quite rapidly without the passage over part (iv). Finally we note that for a sphere our theorem will be an independent (though rather simple) proposition, but that for a multi-surfaced domain S it will be only supplementary to, and not a substitute for the ordinary version, in that the following proposition will be taken as known.

LEMMA 1. In our domain S there exist harmonic functions assuming any prescribed continuous values on the boundary.

2. Dirichlet Integral in Polar Coordinates. The polar coordinates (7) are certain transformations

$$(12) \qquad x_j = r \cdot \omega_j(\theta_1, \ldots, \theta_{k-1}), \; j = 1, \ldots, k$$

$$\omega_1^2 + \ldots \omega_k^2 = 1$$

where $\theta_1, \ldots, \theta_{k-1}$ are Euler angles on (6), say, and for $k = 2$ we have the ordinary transformation

$$(13) \qquad x_1 = r \cos \theta, \qquad x_2 = r \sin \theta.$$

We now take for S the interior of (6) that is

$$(14) \qquad r < 1.$$

For $k = 2$, if we transform $F(x_1, \ldots, x_k)$ into (8), we may set up the expansion

$$(15) \quad f(r;\theta) = (2\pi)^{-1/2} f_0(r) + \sum_{n=1} [f_n(r) \frac{\cos n\theta}{\pi^{1/2}} + g_n(r) \frac{\sin n\theta}{\pi^{1/2}}]$$

in

$$(16) \qquad 0 < r < 1 \quad ,$$

and each $f_n(r)$, $g_n(r)$ is continuous in (16). We can then justify the expansions

$$(17)\quad \frac{\partial f}{\partial r} \sim (2\pi)^{-1/2} f'_0(r) + \sum_{n=1} [f'_n(r) \frac{\cos n\theta}{\pi^{1/2}} + g'_n(r) \frac{\sin n\theta}{\pi^{1/2}}]$$

$$(18)\quad \frac{\partial f}{\partial \theta} \sim \sum_{n=1} [-nf_n(r) \frac{\sin n\theta}{\pi^{1/2}} + ng_n(r) \frac{\cos n\theta}{\pi^{1/2}}] ,$$

and if we substitute them into

$$(19)\quad \int_S [(\frac{\partial F}{\partial x_1})^2 + (\frac{\partial F}{\partial x_2})^2] dx_1 dx_2 = \lim_{\epsilon \to 0, a \to 1} \int_0^{2\pi} d\theta \int_\epsilon^a [(\frac{\partial f}{\partial r})^2 + r^{-2}(\frac{\partial f}{\partial \theta})^2] r\, dr,$$

we obtain

$$(20)\quad ||F||^2 = \int_0^1 [f'_0(r)^2 + \sum_{n=1} (f'^2_n + g'^2_n)] r\, dr + \int_0^1 \sum_{n=1} n^2 (f^2_n + g^2_n) r^{-1}\, dr.$$

For general k we first of all have

$$(21)\quad ||F||^2 = \int_{r=0}^1 (\int_\theta [(\frac{\partial f}{\partial r})^2 + r^{-2} D_\theta(f,f)] dv_\theta) r^{k-1}\, dr$$

where

$$(22)\quad D_\theta(\varphi, \psi) = g^{\alpha\beta} \frac{\partial \varphi}{\partial \theta_\alpha} \frac{\partial \psi}{\partial \theta_\beta}$$

and $g_{\alpha\beta}$ is the metric tensor on (6) as induced by its imbedding in E_k; and

$$(23)\quad dv_\theta = g^{1/2}\, d\theta_1 \ldots d\theta_k.$$

Next we introduce a complete system of spherical harmonics[1]

---

1. See for instance N. Nielsen, "Theorie des Fonctions Metaspheriques." Paris 1911, in connection with E. Heine, "Handbuch der Kugelfunktionen," Band I, 2nd ed., Berlin 1878, pp. 460-461.

$$r^n P_{n,\sigma} (\theta_1, \ldots, \theta_{k-1}) \tag{24}$$

for

$$n = 0,1,2,\ldots ; \qquad \sigma = 1, \ldots, l_n, \tag{25}$$

and we assume that the system

$$\{ P_{n,\sigma} \} \tag{26}$$

is orthonormal on (6), that is

$$\int_\theta P_{m,\rho} \; P_{n,\sigma} \; dv_\theta \; = \; \begin{cases} 1 & \text{if } (m,\rho) = (n,\sigma) \\ 0 & \text{if } (m,\rho) \neq (n,\sigma). \end{cases} \tag{27}$$

If we introduce the Laplacean

$$\Delta_\theta \varphi = \frac{1}{g^{1/2}} \frac{\partial}{\partial \theta_\alpha} \left(g^{1/2} g^{\alpha\beta} \frac{\partial \varphi}{\partial \theta_\beta}\right) \;, \tag{28}$$

then

$$\Delta_\theta P_{n,\sigma} \; = \; -n(n+k-2) P_{n,\sigma} \tag{29}$$

and this implies

$$\int D_\theta(P_{m,\rho} \;,\; P_{n,\sigma}) \, dv_\theta = n(n+k-2) \;\; \delta_m^n \;\; \delta_\rho^\sigma. \tag{30}$$

In analogy to (15) we now set up the expansion

$$f(r;\, \theta_1, \ldots, \theta_{k-1}) \sim \sum\nolimits_{n\sigma} f_{n,\sigma}(r) \; P_{n,\sigma} \; (\theta_1,\ldots,\theta_{k-1}) \;, \tag{31}$$

and it can be shown that it and its formal derivatives

$$\frac{\partial f}{\partial r} \sim \sum\nolimits_{n,\sigma} f'_{n,\sigma} (r) \; P_{n,\sigma} \;\; (\theta_1,\ldots,\theta_{k-1}) \tag{32}$$

(33) $$\frac{\partial f}{\partial \theta_\alpha} \sim \sum_{n,\sigma} f_{n,\sigma}(r) \frac{P_{n,\sigma}}{\partial \theta_\alpha}$$

are boundedly convergent in $(0 <) \epsilon < r < a (< 1)$ and that they may be employed for an evaluation of the right side of (21). With the aid of (27) and (30) we then obtain

(34) $$\|F\|^2 = \sum_{n,\sigma} \int_0^1 [f_n'(r)^2 r^{k-1} + n(n+k-2) f_{n,\sigma}(r)^2 r^{k-3}] \, dr$$

which is a generalization of (20).

3. A One-dimensional Minimum Problem.

THEOREM 2. For $n \geq 0$ and $k \geq 2$, integers or not, if $f(r)$ in (16) is absolutely continuous in the neighborhood of every point and if the norm

(35) $$M^2 = \int_0^1 [f'(r)^2 r^{k-1} + n(n+k-2) f(r)^2 r^{k-3}] \, dr$$

is finite, then

(α) the limit

(36) $$f(1) = \lim_{r \to 1} f(r)$$

exists,

(β) we have

(37) $$n |f(r)|^2 \leq M^2$$

in $0 < r \leq 1$, and

(γ) for $f(1) = 1$ we have

(38) $$M^2 \geq n$$

with equality in (38) holding only for

(39) $$f(r) = r^n$$

PROOF. We make the transformation $r = e^{-t}$ and denote $f(r)$ by $\varphi(t)$ so that

(40) $$M^2 = \int_0^\infty [\varphi'(t)^2 + n(n+k-2)\ \varphi(t)^2]\ e^{-(k-2)t}\ dt\ .$$

From

(41) $$\left(\int_0^1 |\varphi'(t)|\ dt\right)^2 \leqq \int_0^1 \varphi'(t)^2\ dt \leqq e^{k-2} M^2 < \infty$$

it follows that $\varphi(t)$ is absolutely continuous in $0 < t < 1$ so that the limit

(42) $$\varphi(0) = \lim_{t \to 0} \varphi(t)$$

exists, which proves (α).

In proving (β) and (γ) we will first consider only the case $k = 2$ in which case

(43) $$M^2 = \int_0^\infty [\varphi'(t)^2 + n^2 \varphi(t)^2]\ dt$$

and we will also assume $n > 0$, since for $n = 2$ everything can be verified trivially. If we put $\varphi(-t) = \varphi(t)$, $0 < t < \infty$, then $\varphi'(-t) = -\varphi(t)$, and $\varphi(t)$ and $\varphi'(t)$ both belong to $L_2(-\infty, \infty)$. By the theory of Fourier Transforms[2] there exists an even function $\chi(\alpha)$ such that

(44) $$\varphi(t) \sim (2\pi)^{-1/2} \int_{-\infty}^{\infty} e^{it\alpha}\ \chi(\alpha)\ d\alpha$$

$$\varphi'(t) \sim (2\pi)^{-1/2} \int_{-\infty}^{\infty} e^{it\alpha}\ i\alpha\ \chi(\alpha)\ d\alpha\ ,$$

2. See S. Bochner and K. Chandrasekharan, "Fourier Transforms," Princeton 1949, Chapter IV.

and, by Parseval's relation, (40) is then equivalent with

$$(45) \qquad \int_0^\infty \chi(\alpha)\,(\alpha^2 + n^2)\,d\alpha = M^2 .$$

Therefore we have

$$(46) \qquad \left(\int_0^\infty |\chi(\alpha)|\,d\alpha\right)^2 \leqq \int_0 \chi(\alpha)^2(\alpha^2 + n^2)d \quad \int_0 \frac{d\alpha}{\alpha^2+n^2} \leqq M^2 \frac{\pi}{2n} ,$$

and thus

$$(47) \qquad \varphi(t)^2 \leqq \frac{2}{\pi}\left(\int_0^\infty |\chi(\alpha)|\,d\alpha\right)^2 \leqq \frac{M^2}{n} ,$$

which proves part (β), for k = 2. In order to prove (γ) we have to minimize

$$(48) \qquad \int_0^\infty \chi(\alpha)^2\,(\alpha^2 + n^2)\,d\alpha$$

under the condition

$$(49) \qquad \varphi(0) = \left(\frac{2}{\pi}\right)^{1/2} \int_0^\infty \chi(\alpha)\,d\alpha = 1 .$$

Since

$$(50) \qquad \int_0^\infty \chi(\alpha)\,d\alpha \leqq \int_0^\infty |\chi(\alpha)|\,d\alpha ,$$

relations (49) and (46) imply

$$(51) \qquad 1 \leqq M^2 n^{-1} ,$$

which is (38), and equality in (51) can occur if and only if it occurs in (46) and (50); thus, if and only if

$$\chi(\alpha)^2\,(\alpha^2 + n^2) = \text{const.}\,(\alpha^2 + n^2)^{-1}$$

that is

$$\chi(\alpha) = \text{const.}\,(\alpha^2 + n^2)^{-1}$$

and by (44) this is equivalent with $\varphi(t) = \text{const.}\,e^{-nt}$, that is with (39).

For $k > 2$ we put

$$\varphi(t) = \psi(t)\exp\left(\frac{k-2}{2}\,t\right) \tag{52}$$

which implies

$$M^2 = \int_0^\infty [(n + \tfrac{k-2}{2})^2\,\psi(t)^2 + (k-2)\,\psi(t)\,\psi'(t) + \psi'(t)^2]\,dt. \tag{53}$$

Since for $n > 0$ the integrand is a positive definite quadratic form in ψ, ψ', the finiteness of (53) implies the finiteness of the integrals

$$\int_0^\infty \psi(t)^2 dt, \qquad \int_0^\infty \psi'(t)^2\,dt \tag{54}$$

and hence also the existence of

$$\int_0^\infty \psi(t)\,\psi'(t)\,dt. \tag{55}$$

Also[2] the finiteness of the integrals (54) implies the relation

$$\lim_{t\to\infty} \psi(t) = 0\,,$$

and therefore (55) has the value

$$\frac{1}{2}\int_0^\infty d[\psi(t)^2] = -\frac{1}{2}\,\psi(0)^2\,. \tag{56}$$

---

2. See S. Bochner and K. Chandrasekharan, "Fourier Transforms," Princeton 1949, Chapter IV.

Thus (22) implies

$$M^2 + \frac{k-2}{2} \psi(0)^2 = \int_0^\infty [(n + \frac{k-2}{2})^2 \psi(t)^2 + \psi'(t)^2]\, dt \; ; \tag{57}$$

and if we compare this with (43), then by what we have already proven for $k = 2$, we first of all see that for $\psi(0) = \varphi(0) = 1$ we have

$$M^2 + \frac{k-2}{2} \geq n + \frac{k-2}{2}$$

that is (51), with equality occurring only for

$$\psi(t) = \exp\, [-(n + \frac{k-2}{2})\, t] \; ,$$

that is $\varphi(t) = e^{-nt}$, or (39). Furthermore, by (β) for $k = 2$ we obtain

$$(n + \frac{k-2}{2}) \; \psi(t)^2 \leq M^2 + \frac{k-2}{2} \psi(0)^2$$

and hence

$$(n + \frac{k-2}{2}) \sup_t \psi(t)^2 \leq M^2 + \frac{k-2}{2} \sup_t \psi(t)^2$$

or

$$n \sup_t \psi(t)^2 \leq M^2 \; ,$$

and thus

$$nf(r)^2 \leq M^2 r^{\frac{k-2}{2}} \leq M^2 \; ,$$

which completes the proof of the theorem.

4. Interior of a Sphere. If S is the domain

$$x_1^2 + \ldots + x_k^2 < 1 \tag{58}$$

and $||F||$ is finite then we may apply theorem 2 to proving Theorem 1 in the following way. Part (α) implies the existence of the limits

$$f_{n,\sigma}(1) = \lim_{r\to 1} f_{n,\sigma}(r) \tag{59}$$

for all combinations (25), n = 0 included. Next, (β) implies the estimate

$$\sum_{n=1}\left(\sum_{\sigma=1}^{l_n} n \ f_{n,\sigma}(r)^2\right) \leq ||F||^2 \tag{60}$$

uniformly in $0 \leq r \leq 1$. In particular, (59) and (60) imply

$$\sum_{n=0}\sum_{\sigma=1}^{l_n} f_{n,\sigma}(1)^2 < \infty ; \tag{61}$$

and therefore, by the Riesz-Fischer theorem, there exists on (6) an L_2-function, which we will denote by $f(1; \theta_1, \ldots, \theta_{k-1})$, such that

$$f(1; \theta_1, \ldots, \theta_{k-1}) \sim \sum_{n,\sigma} f_{n,\sigma}(1) P_{n,\sigma}(\theta). \tag{62}$$

Also, (59) and (60) imply

$$\lim_{r\to 1} \sum_{n,\sigma} |f_{n,\sigma}(r) - f_{n,\sigma}(1)|^2 = 0, \tag{63}$$

and therefore we have relation (10), which proves part (i) of Theorem 1. We also note, that our boundary function (62) is null if and only if

$$f_{n,\sigma}(1) = 0, \qquad n = 0,1,2, \ldots ; \quad \sigma = 1, \ldots, l_n \tag{64}$$

and that (64) implies not only

$$\int_\theta |f(r;\theta)|^2 \, dv_\theta = \sum_{n,\sigma} |f_{n,\sigma}(r)^2| \to 0, \qquad r \to 1, \tag{65}$$

by (10), but also

$$(66) \qquad \sum_{n,\sigma} n\, f_{n,\sigma}(r)^2 \to 0, \qquad r \to 1,$$

on the basis of (60).

We will next insert a remark, which will be of no particular consequence, about the nature of the subspace L_2^0 of L_2 consisting of boundary functions. For fixed k, if we put $\lambda_n = n(n+k-2)$ then in the series (60) and (66) we may replace the factor n by $\lambda_n^{1/2}$ without any substantial changes of conclusions. Now, by (29), the numbers λ_n are the eigen-values of the positive semi-definite operator $-\Delta_\theta$ on (10), and the numbers $\lambda_n^{1/2}$ thus pertain to its uniquely determined positive semi-definite square root $(-\Delta_\theta)^{1/2}$. Thus L_2^0 is the precise domain of existence of the latter operator, and since the factors $\lambda_n^{1/2}$ may also be inserted in (63) and (66) we not only have convergence in norm of $f(r;\theta)$ towards $f(1;\theta)$, but also of $(-\Delta_\theta)^{1/2} f(r;\theta)$ towards $(-\Delta_\theta)^{1/2} f(1;\theta)$, which then is the precise scope of convergence.

Now, with the numbers (59) we set up the series

$$(67) \qquad \sum_{n,\sigma} r^n f_{n,\sigma}(1)\, P_{n,\sigma}(\theta).$$

Since the system (26) is orthonormal and since the multiplicities l_n are known not to grow faster than a power of n, it follows from (61) that the expansion (67) is absolutely uniformly convergent in every internal sphere $0 < r \leq r_0 < 1$ and thus is the expansion of a well defined harmonic function $H^0(x)$ in (58). Also

$$(68) \qquad ||H^0||^2 = \sum_{n,\sigma} n\, f_{n,\sigma}(1)^2 ,$$

and its boundary function is (62), which proves part (ii) of Theorem 1. Part (iii) follows very directly from (γ) if we compare the expressions (34) and (68), and we are still left with proving the more comprehensive part (iv).

Since $F - H^0$ has a boundary function zero, it suffices to prove that in general we have

$$(69) \qquad (F,H) = 0$$

for any function F with boundary function o, and any harmonic function

$$(70) \qquad H(x) \sim \sum_{n,\sigma} r^n h_{n,\sigma}(1) P_{n,\sigma}(\theta)$$

for which

$$(71) \qquad \sum_{n,\sigma} n\, h_{n,\sigma}(1)^2 < \infty .$$

Now

$$(72) \qquad (F,H) = \lim_{r\to 1} \int_{x_1^2+\dots+x_k^2<r^2} \left(\frac{\partial F}{\partial x_1}\frac{\partial H}{\partial x_1} + \dots \right) dv_x$$

and, due to $\Delta H = 0$, the integral on the right is

$$(73) \qquad \int_\theta f(r,\theta) \frac{\partial h(r,\theta)}{\partial r}\, dv_\theta ,$$

and this has the value

$$(74) \qquad \sum_{n,\sigma} n f_{n,\sigma}(r)\, r^{n-1} h_{n,\sigma}(1) .$$

However

$$(75) \quad \left(\sum_{n,\sigma} |n f_{n,\sigma}(r)\, h_{n,\sigma}(1)|\right)^2 \leqq \sum_{n,\sigma} n f_{n,\sigma}(r)^2 \sum_{n,\sigma} n h_{n,\sigma}(1)^2 ,$$

and (69) follows now from (66) and (71).

5. Other Domains. If B_0 is the outer spherical boundary of S, and if we normalize it to be the surface (6), then there is adjacent to it a shell

$$(76) \qquad r_0 \leqq r < 1$$

which is part of S. By a suitable smoothing process[3] we may construct in (58) a function F(x) which equals F in a sub-shell

3. See S. Bochner, "Remarks on the theorem of Green," Duke Journal 3 (1937), pp. 334-338.

$$(77)\qquad r_1 < r < 1, \qquad r_0 < r_1$$

and has finite norm in (58), and in this way we obtain for F a function (9) with property (10) relative to B_0.

In order to deal with inner boundary spheres we will require a partial analogue to Theorem 2.

LEMMA 2. For $n \geq 0$, $k \geq 2$, if

$$(78)\qquad M^2 = \int_1^\infty [f'(r)^2 r^{k-3} + n(n+k-2) f(r)^2 r^{k-1}] dr$$

is bounded then the limit (36) exists and for $n > 0$ we have

$$(79)\qquad (n + \frac{k-2}{2}) |\dot{f}(r)|^2 \leq M^2 .$$

PROOF. On putting $r = e^t$, $f(r) = \varphi(t)$ we obtain

$$(80)\qquad M^2 = \int_0^\infty [\varphi'(t)^2 + n(n+k-2) \varphi(t)^2] e^{(k-2)t} dt ,$$

and
$$(\int_0^1 |\varphi'(t)| \cdot dt)^2 \leq \int_0^1 \varphi'(t)^2 dt \leq M^2$$

implies first of all (36). Next, for $k = 2$, (80) is the same expression as (40) and thus (79) holds. However, for $k \geq 3$, if we put

$$\psi(t) = \varphi(t) \exp(\frac{k-2}{2} t),$$

we obtain

$$M^2 = \int_0^\infty [(n + \frac{k-2}{2})^2 \psi(t)^2 - (k-2) \psi \psi' + \psi'^2] dt$$

and for $n > 0$ we conclude

$$\int_0^{\infty}[(n + \frac{k-2}{2})^2 \ \psi(t)^2 + \psi'(t)^2]\ dt = M^2 - \frac{k-2}{2}\ \psi(0)^2 \leqq M^2$$

and hence (79).

Suppose now we have normalized an inner sphere B_1 to be the surface (10). Then there exists a layer

(81) $$1 < r < r_1$$

which is a part of S. But then there exists a function $\tilde{F}$ which is defined in the total exterior

(82) $$1 < r < \infty,$$

coincides with F in a sublayer

(83) $$1 < r < r_0, \qquad\qquad r_0 < r_1$$

and vanishes outside a certain large sphere, so as to make the integral

(84) $$\int_{r > 1} [(\frac{\partial \tilde{F}}{\partial x_1})^2 + \dots + (\frac{\partial \tilde{F}}{\partial x_k})^2] = M^2$$

finite. If now we introduce for $\tilde{F}$ the expansion (31), then the latter coincides with F in (83), and since (84) implies

$$\sum_{n,\sigma} [n(n+k-2)\ f_{n,\sigma}\ (r)^2\ r^{k-3} + f'_{n,\sigma}\ (r)^2\ r^{k-1}]\ dr = M^2 < \infty$$

we again conclude from Lemma 2 the existence of a boundary function (9) for which (10) holds, and again the validity of (60) and of similar relations.

Having secured the boundary functions $F_q(\xi)$, we associate with them harmonic functions $H_q(x)$ in the following way. With $F_0(\xi)$ we associate the harmonic functions $H_0(x)$ which is defined, schematically, by the series (67) and thus exists inside the entire sphere bounded by B_0. But with B_1 (and then $B_2, \dots, B_p$) we associate schematically the series

$$\sum_{n,\sigma} r^{-n-k+2} \dot{f}_{n,\sigma}(1)\, P_{n,\sigma}(\theta) \tag{85}$$

and this defines a harmonic function H_1 in the entire exterior of B_1. If now we form the sum $H_0 + H_1 + \ldots + H_p$, then it is defined and has a finite Dirichlet norm in S; and the difference $F - (H_0 + H_1 + \ldots + H_p)$ has boundary functions which are analytic each. If with these boundary functions we construct the harmonic function H_{p+1} according to Lemma 1, then it again has a bounded norm, and the function

$$H^o(x) = H_0 + H_1 + \ldots + H_{p+1}$$

satisfies part (ii) of Theorem 1.

We still have to prove part (iv) and we will again verify (69) under the previous assumptions. We approximate to our domain S by an approximating domain S' which is bounded by spheres $B_0', B_1', \ldots, B_p'$ such that each B_q' is concentric with B_q, and we will prove that

$$\int_{S'} \left(\frac{\partial F}{\partial x_1} \frac{\partial H}{\partial x_1} + \ldots + \frac{\partial F}{\partial x_k} \frac{\partial H}{\partial x_k} \right) dv_x \tag{86}$$

tends to 0 as $S' \to S$. Now, the integral (86) is the sum over q of the boundary integrals

$$\int_{B_q'} F \frac{\partial H}{\partial \nu} \, d\omega_\xi \, , \tag{87}$$

and we claim that each of these converges to 0 as $B_q' \to B_q$. Consider for instance the outer boundary B_0 - the others can be dealt with in a similar way - and assume again that it is the surface (6). In the adjacent layer (76) we can write $H = H^1 + H^2$ where H^2 exists for $r \geq r_0$ and H^1 exists and is of finite norm in (58), and it suffices to show that the two integrals

$$\int f(r,\theta) \frac{\partial H^1}{\partial r} \, d\omega_\theta \qquad \int f(r,\theta) \frac{\partial H^2}{\partial r} \, d\omega_\theta$$

separately tend to 0 as $r \to 1$. For the first integral this has already been proven in Section 4, and for the second integral the proof is even simpler. The square of its absolute value is

$$\leq \int f(r,\theta)^2 \, d\omega_\theta \cdot \int \left| \frac{\partial H^2}{\partial r} \right|^2 d\omega_\theta ,$$

but the second factor remains bounded as $r \to 1$ and the first tends to 0 by relation (65), and thus the proof of our theorem is completed.

6. General Remarks on Dirichlet Principle. If in an arbitrary domain S of the Euclidean E_k we consider the family of harmonic functions $h(x_1,\ldots,x_k)$, then it follows from the Poisson integral formula, as applied to a circular neighborhood of any point, that a bound of $|h(x)|$ in a compact subset S' of S also majorizes each partial derivative

$$\frac{\partial^{r_1 + \ldots + r_k} f}{\partial x_1^{r_1} \ldots x_k^{r_k}}$$

in any compact subset S" of the interior of S'; so that, in particular, any sequence of harmonic functions which are equally bounded in every S' contains a subsequence which is uniformly convergent, with every partial derivative, in every S'. But more is true. The Poisson formula gives in particular the volume average

$$h(x) = \frac{1}{m(u)} \int_{u(x^0,r)} h(x+y) \, dv_y$$

from which we infer by Hölder-Minkowski that

$$|h(x)|^p \leq \frac{1}{m(u)} \int_u |h(x+y)|^p \, dv_y \quad . \tag{88}$$

Therefore, uniform bounds for the volume integrals

$$\int_{S'} |h(y)| \, dv_y ,$$

and the more so for the integrals

$$\left(\int_{S'} |h(y)|^p \, dv_y \right)^{1/p},$$

give uniform bounds for h(x) itself and for its partial derivatives, and imply the sequential compactness previously described. Also, if a sequence of harmonic functions is convergent in L_p-norm over S', it also converges uniformly in S", so that the L_p-limit is also a uniform limit, and hence again a harmonic function.

We now take k real (non-harmonic) functions $f_1(x), \ldots, f_k(x)$ which belong to L_2 over S, and are determined only up to sets of Lebesque measure 0, and we associate with the vector

$$f = (f_1, \ldots, f_k) \tag{90}$$

the Dirichlet-Hilbert norm

$$||f|| = \left(\int_S (f_1^2 + \ldots + f_k^2)\, dv\right)^{1/2} \tag{91}$$

which makes the space of such vectors into a Hilbert space with the inner product

$$(f,g) = \int_S (f_1 g_1 + \ldots + f_k g_k)\, dv . \tag{92}$$

This Hilbert space will be denoted by Λ.

We now consider the additive subspace L^0 consisting of such elements

$$H = (h_1, \ldots, h_k)$$

of Λ as are gradient fields of harmonic functions

$$h_j = \frac{\partial H}{\partial x_j}, \qquad j = 1, \ldots, k,$$

and we claim that L^0 is complete in itself and thus a closed subspace of Λ. In fact, if a sequence of gradient fields

$$h_j^{(n)} = \frac{\partial H^{(n)}}{\partial x_j}$$

is convergent in the norm (91), then every component is so convergent, and hence the components are convergent uniformly. If therefore the additive constant in $H^{(n)}(x)$ is so normalized as to make all $H^n(x^0) = 0$ at a joint point x^0, then these functions are likewise uniformly continuous, the limit being again harmonic, and the limiting vector field being its gradient field.

Now, in any Hilbert space Λ, if f is an arbitrary fixed element of it, and h is a variable element of a closed subspace L^0 then $||f - h||$, for h in L^0, attains its minimum, and the minimizing element h^0 is unique, and it is characterized by the fact that we have $(f - h^0, h) = 0$ for all h in L^0. All of which adds up to being an "abstract" version of Dirichlet's Principle without any reference to boundary values.

We note that the results just stated apply to any element f of Λ and not only to a continuous gradient field

$$(93) \qquad f_j = \frac{\partial F}{\partial x_j}, \qquad j = 1, \ldots, k,$$

say. But now take a sequence of such gradient fields

$$(94) \qquad f_j^{(n)} = \frac{\partial F^{(n)}}{\partial x_j}$$

in Λ and make the following two-part assumption; the sequence of elements $\{f_j^{(n)}\}$ converge in Λ towards some element $\{f_j\}$ in norm, and the functions $F^{(n)}(x)$ themselves also converge towards some continuous function F(x), uniformly in every S', say. It can then be shown that this function is then independent of the special sequence of gradient fields (94) thus converging towards f, and that it is what we have called a "weak solution"[4] of the system (93); the meaning of this last statement being that for any functions $\varphi_j(x)$ with continuous derivatives in all of S, but vanishing outside some neighbordood in S each, we have the DuBois Reymond relations

$$\int_S \left(F \frac{\partial \varphi_j}{\partial x_j} + f_j \varphi_j\right) dv = 0.$$

It can further be shown that our theorem 1 fully applies to a function F(x) of this description, and this probably covers every type of "piecewise differentiable" functions to which our theorem would be extensible by interpreting the concept of a derivative in the literal manner.

4. S. Bochner, "Linear partial differential equations with constant coefficients," Annals of Math. 47 (1946), pp. 202-212. Compare also Bochner and W. T. Martin, Several complex variables, Princeton 1948, p. 158 ff.

Returning to the Principle itself, we now define a Banach space Λ_p by using the norm

$$(95) \qquad ||f||_p \doteq \left(\int_S (|f_1|^p + \dots + |f_k|^p)\, dv \right)^{1/p}$$

instead of (91), for $p \geqq 1$. The subspace L_p^0 of gradient fields of harmonic functions is again a closed subspace, and we claim that

$$\lambda_f = \inf_{h \in L_p^0} ||f - h||$$

is again $||f - h^0||$ for some h^0. In fact if we pick a minimizing sequence $h^{(n)}$ such that

$$||f - h^{(n)}|| \to \lambda_f$$

then a subsequence k converges towards a gradient field h^0 for which we have

$$\left(\int_{S''} (|f_1 - h_1^0|^p + \dots + |f_k - h_k^0|^p)\, dv \right)^{1/p} \leqq \lambda_f$$

for each compact S" in S, and hence by a limiting process for S in lieu of S". We thus obtain $||f - h^0|| \leqq \lambda_f$, and therefore finally

$$||f - h^0|| = \lambda_f .$$

If there are two minimizing functions h^1, h^2 then this implies

$$||f - \frac{h^1 + h^2}{2}|| = ||\frac{f - h^1}{2}|| + ||\frac{f - h^2}{2}|| ,$$

but for $p > 1$ the unit sphere in Λ_p is strictly convex,[5] and the last relation implies $h^1 = h^2$, so that we again have uniqueness. But for $p = 1$ we can imply $h_j^1 = h_j^2$ only on such subsets S^0, on which neither $f_j - h_j^1 = 0$ nor $f_j - h_j^2 = 0$. If now the components f_j are all analytic functions in the

5. Compare J. A. Clarkson, "Uniformly convex spaces," Trans. Amer. Math. Soc. 40 (1936), pp. 396-414.

real variables $x_1, \ldots, x_k$ then we again have $h^1 = h^2$ throughout.[6]

Finally we note that these Λ_p-statements also hold for the modified norm

$$\left(\int_S (f_1{}^2 + \ldots + f_k{}^2)^{p/2}\, dv \right)^{1/p}$$

which is structurally closer to (91) than (95) is.

7. Remarks on the Problem of Plateau. For $k = 2$ it was shown by Douglas[7] that the Dirichlet integral (20) has apart from a factor the value

$$(95) \qquad \int_0^{2\pi} \int_0^{2\pi} \frac{(f(\theta) - f(\varphi))^2}{\sin^2 \frac{\theta - \varphi}{2}}\, d\theta\, d\varphi \ ,$$

where $f(\theta)$ is our boundary function of $F(x)$. This expression is valid for non-continuous functions as well, and the proof of Douglas is then also valid. His solution of the problem of Plateau consisted in showing that if a system of continuous functions $f_1, \ldots, f_n$ with finite values (95) constitute a curve without multiple points, then the value of

$$\int_0^{2\pi} \int_0^{2\pi} \frac{\sum_{i=1}^{n} (f_i(\lambda(\theta)) - f_i(\lambda(\varphi)))^2}{\sin^2 \frac{\theta - \varphi}{2}}\, d\theta\, d\varphi$$

for all one-one continuous transformations $\theta^1 = \lambda(\theta)$ of the circle into itself attains its lowest possible value for at least one such transformation $\lambda^*(\theta)$ likewise continuous.

Any such transformation carries measurable functions into measurable functions, and we wish to point out that Douglas' argument retains its validity for non-continuous functions $f_i(\theta)$ under the following additional assumptions. The functions $f_i(\theta)$ are bounded, and what is decisive, corresponding to any $\epsilon > 0$ there exists a $\delta > 0$ such that for $|\theta - \varphi| \geq \epsilon$ we have

$$(f_i(\theta) - f_i(\varphi))^2 \geq \delta(\epsilon),$$

and this is a rather precise way of demanding that the "curve" have no multiple points, without invoking continuity.

---

6. Compare for this Bochner and W. T. Martin, l.c., p. 118.

7. J. Douglas, "Solution of the problem of Plateau," Trans. Amer. Math. Soc. 33 (1931), pp. 263-321.

III. THE FRÉCHET VARIATION AND PRINGSHEIM CONVERGENCE OF DOUBLE FOURIER SERIES

By Marston Morse and William Transue[1]

1. Introduction. Tests for the Pringsheim convergence of the double Fourier series of an integrable even-even function f with values f(u,v) and period 2π in each variable have been given by Hardy (1), W. H. Young, Tonelli, Gergen (1) and others. The series is tested for convergence to s at the origin and one sets $\varphi(u,v) = f(u,v) - s$. Practically all of the tests previously given use directly or indirectly the Vitali variation either of a function or of an indefinite integral. To recognize this fact one must recall that if g is integrable over an interval Q the V-variation [V=Vitali] $V(Q,\bar{g})$ of the indefinite integral $\bar{g}$ of g over Q is

$$V(Q,\bar{g}) = \int_Q\int |g(s,t)|\,dsdt. \tag{1.1}$$

For example the generalized Dini condition [see Gergen (1)] may be stated in the form

$$V\,[I, \overline{\varphi/uv}] < \infty \tag{1.2}$$

where I is the interval $(0,\pi] \times (0,\pi]$.

That this almost universal use of the Vitali variation in the formulation of tests is not altogether natural and is unnecessarily restrictive appeared to the authors as a consequence of their study of bilinear functionals over the space of Cartesian products $C \times C$. Here C is the space so designated by Banach. For the Dirichlet integral used in representing the partial sums S_{mn} of the Fourier series appeared as a bilinear functional in which one substituted elements f_n and f_m in the space $C(0,\pi]$ with

1. The contribution of Dr. Transue to this paper is in partial fulfillment of a contract between the Office of Naval Research and the Institute for Advanced Study.

interval $(0, \pi]$, where

$$f_n(u) = \frac{\sin(n + \frac{1}{2})u}{\sin \frac{u}{2}} \qquad f_m(v) = \frac{\sin(m + \frac{1}{2})v}{\sin \frac{v}{2}}$$

For most purposes the most natural technical instrument for studying bilinear functionals H on C X C is the F-variation [F=Fréchet] applied to the distribution function associated with H.

The application of this general concept to the study of Fourier series became possible only because in other connections (namely the variational theory of general quadratic functionals) a technique had been developed. With this start each of the tests considered by Gergen (1) (with the exception of that originating with Tonelli) was given a form in which the implicit use of the V-variation was explicitly replaced by a use of the F-variation. Tests designated by Gergen (1) by symbols

$$(1.3) \qquad (J_H)\ (V_Y)\ (D_Y)\ (Y)\ (Y_P)\ (L_1)\ (L_2)\ (L_P)\ (L_R)$$

were so modified and denoted by the same symbol with the parenthesis removed. For convenience we term the tests (X) in (1.3) <u>classical</u> and our new tests X, MT-tests. To give an example our Dini test D_Y, replacing (D_Y) as given by (1.2), is merely

$$(1.4) \qquad P(I, \overline{\varphi/uv}) < \infty \qquad \{I = (0, \pi] \times (0, \pi]\}$$

where $P(I,\bar{g})$ is the F-variation of the function $\bar{g}$ over the interval I, and $\bar{g}$ is any indefinite integral of g over I. Our new tests are explicitly defined in §2 together with the classical tests (1.3).

We show first that each of the MT-tests is sufficient for the Pringsheim convergence of the Fourier series at the origin to s. The greatest interest in this new departure is of course a comparison of the classes K^X and $K^{(X)}$ of functions φ respectively satisfying an MT-test X and the corresponding classical test (X). It should be recalled that the test (L_R) is due to Gergen who has shown that the class $K^{(L_R)}$ includes each other classical class $K^{(X)}$. However no proof has apparently been given by Gergen or others that the other classes of Lebesgue type

$$(1.5) \qquad K^{(L_P)}, \quad K^{(L_1)}, \quad K^{(L_2)}$$

are proper (i.e. smaller) subclasses of $K^{(L_R)}$. With this background our first comparison theorem is somewhat surprising.

THEOREM 1. 1. The Gergen class $K^{(L_R)}$ includes none of the MT-classes

$$K^{L_R},\ K^{L_P},\ K^{L_1},\ K^{L_2},\ K^{V_Y},\ K^{D_Y}. \tag{1.6}$$

The class $K^{(Y_P)}$ includes none of the MT-classes. Each MT-class K^X includes the corresponding classical class $K^{(X)}$ as a proper subclass. Thus the MT-class K^{L_R} includes each classical class as a proper subclass.

In proving the non-inclusion relations it was necessary to depart from the classical examples, due largely to Hardy, and from the product functions with values $\psi_1(u)\,\psi_2(v)$, and use (apparently for the first time) truly two dimensional examples. A minor but interesting gap[2] in the 1-dimensional theory dating from Pollard[3] 1927 has been filled by the authors on showing that for 1-dimensional tests in the notation of Gergen (2) p. 255, $(Y_P) \Rightarrow (L_2)$. See MT (10).

The relation between the MT-tests themselves remain substantially the same as those between the classical tests as established by Gergen (1). In Figure 1 part of these relations are represented. Two tests X and Y are termed <u>incomparable</u> if neither of the classes K^X or K^Y includes the other. In Figure 1 the class K^Y is represented by a semi-circular region bounded by the semi-circle bearing the symbol Y. The intersection of two semi-circular regions indicates the incomparability of the corresponding tests. Inclusion of a region in another indicates inclusion of the corresponding classes.

In Figure 2 a full vector leading from X to Y indicates that $X \Rightarrow Y$ in the sense that a function φ satisfying conditions X also satisfies conditions Y. The dotted vectors indicate an implication whose truth or falsity remains unestablished. Apart from the dotted vectors the diagram is <u>complete</u> in that any additional full vector drawn between two vertices,

---

2. In the notation of Gergen (2)

$$(Y_P) = (Y') + (C_1).$$

Pollard shows §(9) that $(Y_P) \Rightarrow (L_P)$ (Gergen's notation). Our result $(Y_P) \Rightarrow (L_2)$ implies this theorem of Pollard since $(L_2) \Rightarrow (L_P)$. See diagram Gergen (2) p. 255 and footnote.

3. Referring to the theorem that (Y_P) implies convergence of the Fourier series Pollard writes: "There appears to be no simple way of deducing it from the commonly accepted form of Lesbesgue's criterion." In establishing the relation $(Y_P) \Rightarrow (L_2)$ we have however given such a deduction.

but not logically implied by Figure 2 or indicated as doubtful, would have a false implication. For example, Figure 2 indicates that conditions Y_P are not implied by conditions D_Y, V_Y, L_1, L_P, L_R.

In terms of φ and with $x \geq 0 \quad y \geq 0$ set

$$\hat{\varphi}(x,y) = \int_0^x \int_0^y \varphi(s,t)\, ds\, dt \ .$$

Let I_{xy} denote the interval $(0,x] \times (0,y]$. Conditions C^0 and C^O (weak versions of classical conditions) are as follows.

$$C^0: \quad P[I_{xy}, \hat{\varphi}] = o(xy) \qquad (x > 0,\ y > 0)$$

$$C^O: \quad P[I_{xy}, \hat{\varphi}] = O(xy) \qquad (x > 0,\ y > 0) \ .$$

The equivalences

$$L_1 \equiv L_2 \equiv L_P + C^0 \equiv L_R + C^0 \ ; \quad L_P \equiv L_R + C^O$$

parallel equivalences found by Gergen in the V-theory.

We have not generalized the Tonelli Test [Gergen's extension is denoted by (J_R)] since (J_R) does not use the V-variation.

However our class K^{L_R} includes the Tonelli class as a proper subclass since

$$K^{L_R} \supset K^{(L_R)} \supset K^{(J_R)}$$

and since the first inclusion (due to us) is proper. The second inclusion is due to Gergen (1).

In preparing for this paper new aspects of integration theory were discovered. In particular we find that our FL-integrals are a subclass of Harnack-Lebesgue integrals with special properties not shared by HL-integrals in general. See MT (11).

2. The tests defined.

Limit notation. This notation follows Landau and Gergen (1). Given functions h and ψ with values $h(x,y)$ and $\psi(x,y)$ defined for all sufficiently small positive values x and y write

$$h(x,y) = o|\psi(x,y)| \tag{2.1}$$

if corresponding to each $\epsilon > 0$ there exists a $\delta(\epsilon) > 0$ such that

$$|h(x,y)| \leqq \epsilon|\psi(x,y)| \qquad [0 < x < \delta(\epsilon)]\ [0 < y < \delta(\epsilon)] \tag{2.2}$$

Similarly write

$$h(x,y) = O|\psi(x,y)| \tag{2.3}$$

if (2.2) holds for some $\epsilon > 0$ and all sufficiently small positive x and y. Let k be any positive number exceeding some constant and h_k and ψ_k be any two functions of the type of h and ψ. We write

$$h_k(x,y) = \bar{o}|\psi_k(x,y)| \tag{2.4}$$

if corresponding to each $\epsilon > 0$ there exists a $k_\epsilon > 0$ and then a $\delta(k, \epsilon) > 0$ such that

$$|h_k(x,y)| \leqq \epsilon|\psi_k(x,y)| \tag{2.5}$$

for

$$0 < x < \delta(k, \epsilon), \quad 0 < y < \delta(k, \epsilon) \qquad (k \geqq k_\epsilon). \tag{2.6}$$

Finally write

$$h_k(x,y) = \bar{O}|\psi_k(x,y)| \tag{2.7}$$

if corresponding to some $\epsilon > 0$ there exists a k_0 and $\delta(k)$ such that (2.5) holds for some constant ϵ for $0 < x < \delta(k)$, $0 < y < \delta(k)$, and $k > k_0$. If for fixed (x,y), $h_k(x,y)$ <u>decreases</u> <u>monotonically</u> as k increases the preceding condition can be stated in terms of a δ_0 independent of k.

Conditions (2.1), (2.3), (2.4) and (2.7) respectively imply that

$$h(x,y) = \psi(x,y)o(1), \qquad h(x,y) = \psi(x,y)O(1)$$

$$h_k(x,y) = \psi_k(x,y)\bar{o}(1) \qquad h_k(x,y) = \psi_k(x,y)\bar{O}(1).$$

The relations

$$h(x,y) = o(1),\ h(x,y) = O(1),\ h_k(x,y) = \bar{o}(1),\ h_k(x,y) = \bar{O}(1)$$

respectively imply that

$$h(kx,ky) = \bar{o}(1) \qquad h(kx,ky) = \bar{O}(1)$$

$$h_k(kx,ky) = \bar{o}(1) \qquad h_k(kx,ky) = \bar{O}(1)$$

The relation $h(x,y) = o(1)$ implies the relations

$$\begin{aligned} \lambda(k)h(kx,ky) &= \bar{o}(1) \\ \lambda(k)h(x,\ ky) &= \bar{o}(1) \qquad (2.8) \\ \lambda(k)h(kx,\ y) &= \bar{o}(1) \end{aligned}$$

for arbitrary $\lambda(k)$, etc.

Intervals. An open interval in the 2-space R_2 of points (u,v) of the form of a Cartesian product $(a,a') \times (b,b')$ of two open intervals will be denoted by $\begin{pmatrix} a',b' \\ a\ ,b \end{pmatrix}$. The corresponding closed interval will be denoted by $\begin{bmatrix} a',b' \\ a\ ,b \end{bmatrix}$. More generally we shall admit 2-intervals of the form $Q = U \times V$, where U is any one of the intervals

$$(a,a') \quad [a,a'] \quad (a,a'] \quad [a,a')$$

and V any one of the corresponding intervals in which b and b' replace a and a'.

The Fréchet variation. Let g map a general interval Q into R_1, the axis of reals. If Q is closed the F-variation $P(Q,g)$ of g over Q is well definied. See Fréchet or MT (1). If Q is not closed one defines $P(Q,g)$ by the condition

$$P(Q,g) = \sup_J P(J,g)$$

taking the sup, over all closed intervals $J \subset Q$. If $Q = \begin{bmatrix} a',b' \\ a\ ,b \end{bmatrix}$ or $\begin{pmatrix} a',b' \\ a\ ,b \end{pmatrix}$ we shall write respectively

$$\begin{cases} P(Q,g) = P \begin{bmatrix} a',b' \\ a\ ,b \end{bmatrix} (g) \\ V(Q,g) = V \begin{bmatrix} a',b' \\ a\ ,b \end{bmatrix} (g) \end{cases} \qquad \begin{cases} P(Q,g) = P \begin{pmatrix} a',b' \\ a\ ,b \end{pmatrix} (g) \\ V(Q,g) = V \begin{pmatrix} a',b' \\ a\ ,b \end{pmatrix} (g) . \end{cases}$$

The class $\hat{F}(Q)$. Let K be a straight line segment parallel to an edge of Q and in Q. The segment K need be neither closed nor open. If g is defined over Q, the function $g|K$ defined by g over K maps K into R_1 and the total variation $T(K,g)$ of $g|K$ over K is well defined and possibly infinite. If K and K' are two sections of Q parallel to the same edge of Q

$$(2.9) \qquad T(K,g) \leq T(K',g) + P(Q,g) .$$

See Fréchet or equation (3.0) MT (5). We say that g is in $\hat{F}(Q)$ or satisfies $\hat{F}$ over Q if $P(Q,g) < \infty$ and if for at least one section K of Q parallel to an arbitrary edge of Q, $T(K,g) < \infty$. If g is in $\hat{F}(Q)$ g is bounded and measurable over Q. See MT (6) §2.

The class $\hat{V}(Q)$. This class is defined as was $\hat{F}(Q)$ replacing the condition $P(Q,g) < \infty$ by the condition $V(P,g) < \infty$ in its definition.

The class FL(Q). The function g will be said to be _weakly_ in L over Q if g is in L over every closed subinterval of Q. If Q is closed g is in L(Q) if weakly in L over Q. If g is _given as weakly_ in L _over_ Q we shall set

$$(2.10) \qquad \bar{g}(u,v) = \int_p^u \int_q^v g(s,t)\, ds\, dt \qquad [(p,q) \in Q]$$

terming (p,q) the vertex of $\bar{g}$. If

$$(2.11) \qquad P(Q,\bar{g}) < \infty$$

we shall say that g is in FL(Q). As seen in §6 of MT(11) the value of $P(Q,\bar{g})$ is independent of the choice of the vertex $(p,q) \in Q$. _If_ g _is in_ FL(Q), $\bar{g}$ _admits a continuous extension_[4] _over_ $\bar{Q}$ _called an_ FL-_integral of_ g over $\bar{Q}$. See §6 MT (11). In general a function $g \in FL(Q)$ is not in L(Q). See Th. 7.1 MP (7). If g is in L over $\begin{bmatrix} \pi, \pi \\ 0\ , 0 \end{bmatrix}$ we set

4. The continuous extension of an FL-integral $\bar{g}$ over $\bar{Q}$ will again be denoted by $\bar{g}$.

$$(2.12)\qquad \int_0^u \int_0^v g(s,t)\,ds\,dt = \hat{g}(u,v) \qquad \left\{ (u,v) \in \begin{bmatrix} \pi, & \pi \\ 0, & 0 \end{bmatrix} \right\}.$$

We are concerned with the convergence of the Fourier series of f at the origin to a constant s and set

$$(2.13)\qquad f(u,v) - s = \varphi(u,v)$$

In defining the various tests use will be made of the following conditions.

$$(C_0)\qquad \varphi(x,y) = o(1)$$

$$(C_1)\qquad \hat{\varphi}(x,y) = o(xy)$$

$$(2.14)\qquad C^o \qquad P \begin{bmatrix} x, & y \\ 0, & 0 \end{bmatrix} (\hat{\varphi}) = o(xy)$$

$$C^O \qquad P \begin{bmatrix} x, & y \\ 0, & 0 \end{bmatrix} (\hat{\varphi}) = O(xy)$$

$$C^A \qquad P \begin{bmatrix} x, & y \\ 0, & 0 \end{bmatrix} (\hat{\varphi}) \leq A\,xy \qquad (0 < x \leq \pi)\ (0 < y \leq \pi)$$

where A is a constant. In Gergen C^o, C^O and C^A are replaced by the conditions

$$(2.15)'\qquad (C^o)\qquad V \begin{bmatrix} x, & y \\ 0, & 0 \end{bmatrix} (\hat{\varphi}) = o(xy)$$

$$(2.15)''\qquad (C^O)\qquad V \begin{bmatrix} x, & y \\ 0, & 0 \end{bmatrix} (\hat{\varphi}) = O(xy)$$

$$(2.15)'''\qquad (C^A)\qquad V \begin{bmatrix} x, & y \\ 0, & 0 \end{bmatrix} (\hat{\varphi}) \leq A\,xy$$

The conditions (2.15) are easily seen to be more restrictive than the corresponding conditions C^o, C^O, C^A. The left numbers of (2.15) are ordinarily written in the form

$$(2.16)\qquad \int_0^x \int_0^y |\varphi(s,t)|\,ds\,dt .$$

We write these conditions in the form (2.15) to make the relation between the conditions using V and those using P clear.

Our tests are usually defined by two or more conditions on φ. If A and B are two such conditions A + B shall denote the set of conditions A and B. We shall refer to formal differences

$$(2.17)' \qquad \Delta_x g(u,v) = g(u+x,v) - g(u,v) \qquad (x > 0)$$

$$(2.17)'' \qquad \Delta_y g(u,v) = g(u,v+y) - g(u,v) \qquad (y > 0)$$

$$(2.17)''' \qquad \Delta_{xy} g(u,v) = g(u+x,v+y) - g(u+x,v) - g(u,v+y) + g(u,v) .$$

The functions defined by these values will be respectively denoted (with Gergen) by

$$(2.18) \qquad \Delta_x g \qquad \Delta_y g \qquad \Delta_{xy} g$$

In addition to coordinates (u,v), coordinates (s,t) will be used. The functions

$$(2.19) \qquad \Delta_s g \qquad \Delta_t g \qquad \Delta_{st} g$$

are then defined as in (2.17) replacing x and y by s and t. The objections to this somewhat imperfect notation seem outweighed by its convenience. We also make an anomalous use of g/u, g/v, g/uv, g uv etc. as designating functions with values

$$\frac{g(u,v)}{u} , \frac{g(u,v)}{v} , \frac{g(u,v)}{uv} , uv\, g(u,v) \text{ etc.}$$

respectively. The following conditions are components of our tests.[5]

$$Y' \qquad P \begin{bmatrix} x , y \\ 0 , 0 \end{bmatrix} \left\{ uv\, \varphi \right\} \leq A\, xy \qquad (0 < x \leq \pi)\ (0 < y \leq \pi)$$

---

5. The interval $\begin{bmatrix} x , & y \\ 0+, & 0+ \end{bmatrix}$ is the Cartesian product $(0,x] \times (0,y]$.

$$\begin{cases} L_1' & P\begin{bmatrix}\pi-x, \pi-y\\ x \quad , y\end{bmatrix}\left\{\overline{\frac{1}{uv}\Delta_{xy}\varphi}\right\} = o(1) \\ L_1'' & P\begin{bmatrix}x, \pi-y\\ 0, y\end{bmatrix}\left\{\overline{\frac{1}{v}\Delta_{y}\varphi}\right\} = o(x); \quad P\begin{bmatrix}\pi-x, y\\ x \quad , 0\end{bmatrix}\left\{\overline{\frac{1}{u}\Delta_{x}\varphi}\right\} = o(y)\end{cases}$$

$$\begin{cases} L_2' & P\begin{bmatrix}\pi-x, \pi-y\\ x \quad , y\end{bmatrix}\left\{\overline{\Delta_{xy}\left(\frac{\varphi}{uv}\right)}\right\} = o(1) \\ L_2'' & P\begin{bmatrix}x, \pi-y\\ 0, y\end{bmatrix}\left\{\overline{\Delta_{y}\left(\frac{\varphi}{v}\right)}\right\} = o(x); \quad P\begin{bmatrix}\pi-x, y\\ x \quad , 0\end{bmatrix}\left\{\overline{\Delta_{x}\left(\frac{\varphi}{u}\right)}\right\} = o(y)\end{cases}$$

$$\begin{cases} L_P' & P\begin{bmatrix}\pi-x, \pi-y\\ kx \quad , ky\end{bmatrix}\left\{\overline{\Delta_{xy}\left(\frac{\varphi}{uv}\right)}\right\} = \bar{o}(1) \\ L_P'' & P\begin{bmatrix}x, \pi-y\\ 0, ky\end{bmatrix}\left\{\overline{\Delta_{y}\left(\frac{\varphi}{v}\right)}\right\} = \bar{o}(x); \quad P\begin{bmatrix}\pi-x, y\\ kx \quad , 0\end{bmatrix}\left\{\overline{\Delta_{x}\left(\frac{\varphi}{u}\right)}\right\} = \bar{o}(y)\end{cases}$$

$$\begin{cases} L_R' & P\begin{bmatrix}\pi-x, \pi-y\\ kx \quad , ky\end{bmatrix}\left\{\overline{\frac{1}{uv}\Delta_{xy}\varphi}\right\} = \bar{o}(1) \\ L_R'' & P\begin{bmatrix}x, \pi-y\\ 0, ky\end{bmatrix}\left\{\overline{\frac{1}{v}\Delta_{y}\varphi}\right\} = \bar{o}(x); \quad P\begin{bmatrix}\pi-x, y\\ kx \quad , 0\end{bmatrix}\left\{\overline{\frac{1}{u}\Delta_{x}\varphi}\right\} = \bar{o}(y)\end{cases}$$

MT-<u>Conditions</u>. We understand that the conditions

$$(Y') \quad (L_1') \quad (L_1'') \quad (L_2') \quad (L_2'') \quad (L_P') \quad (L_P'') \quad (L_R') \quad (L_R'')$$

are as defined by Gergen (1). These conditions differ from our conditions without parenthesis as defined above only in the use of V in place of P. We refer to these tests as classical tests and to our new tests as MT-tests. Set $I = \begin{bmatrix}\pi, \pi\\ 0+, 0+\end{bmatrix}$. The MT-tests and corresponding classical tests on $\varphi = f - s$ are as follows.

<u>MT-tests</u>		<u>Classical tests</u>	
J_H:	$\hat{F}(I) + (C_0)$	(J_H):	$\hat{V}(I) + (C_0)$
V_Y:	$\hat{\varphi}/uv$ satisfies J_H	(V_Y):	$\hat{\varphi}/uv$ satisfies (J_H)

MT-tests	Classical tests
D_Y: $P(I, \overline{\varphi/uv}) < \infty$	(D_Y): $V(I, \overline{\varphi/uv}) < \infty$
$Y = Y' + (C_0)$	$(Y) = (Y') + (C_0)$
$Y_P = Y' + (C_1)$	$(Y_P) = (Y') + (C_1)$
$L_1 = L_1' + L_1'' + C^0$	$(L_1) = (L_1') + (L_1'') + (C^0)$
$L_2 = L_2' + L_2''$	$(L_2) = (L_2') + (L_2'')$
$L_P = L_P' + L_P'' + (C_1)$	$(L_P) = (L_P') + (L_P'') + (C_1)$
$L_R = L_R' + L_R'' + (C_1)$	$(L_R) = (L_R') + (L_R'') + (C_1)$

The above conditions (C_0) and (C_1) involve neither the V- or the F-variation. The notation (C_0) (C_1) is old, the notation

$$C^{\circ} \quad C^{0} \quad C^{A} \qquad (C^{\circ}) \quad (C^{0}) \quad (C^{A})$$

new.

Each of the above tests involves the proposed limit s, since $\varphi(u,v) = f(u,v) - s$. In connection with V_Y set

$$\theta(u,v) = \frac{\hat{f}(u,v)}{uv} \qquad \{(u,v) \in I\}$$

Where $F(I)$ and $V(I)$ are satisfied by φ we shall see that the limits $f(0+,0+)$ and $\theta(0+,0+)$ of f and θ from the open first quadrants exist. In these cases the conditon (C_0) on φ implies that $s = f(0+,0+)$ and $s = \theta(0+,0+)$ respectively. With this understood Gergen {and his predecessors in the case of (J_H), (V_Y) and (D_Y)} have shown that the "classical" tests on φ imply the convergence of S_{mn} to s at the origin. We shall establish the corresponding theorem for the MT-tests. We first show that L_R is sufficient for the convergence of S_{mn} to s, and then show that each other MT-test X implies L_R.

3. <u>The conditions</u> L_R <u>proved sufficient</u>. The axes in the (u,v)- plane are lines of singularity of the integrands appearing in the conditions L_R' and L_R''. Because of this it is useful to partition the square $I = \begin{bmatrix}\pi, \pi\\ 0, 0\end{bmatrix}$ by straight lines $u = ax$, $v = by$, $u = \pi - ax$, $v = \pi - by$, where ax and by are small and positive, and to consider separately the nine different subrectangles into which I is divided by these lines. The lines $u = \pi - ax$ $v = \pi - by$ separating off the upper and right edges of Q correspond to the special limits in the conditions, L_R' and L_R'', which in turn are necessitated by the use of the differences $\Delta_x f$ $\Delta_y f$ and $\Delta_{xy} f$ in these conditions. Lemmas 3.2 to 3.7 are concerned with these nine subrectangles of I and show that the contribution of each such subrectangle to the Dirichlet integral

$$(3.1) \qquad D_\varphi(x,y) = \int_I\int \varphi(u,v)\, K(u,v;\, x,y)\, du\, dv$$

is o(1). Here

$$K(u,v;\, x,y) = K^{x,y}(u,v) = \sin\frac{\pi u}{x}\csc\frac{u}{2}\sin\frac{\pi v}{y}\csc\frac{v}{2}$$

As is well known the problem of proving that S_{mn} converges to s may be solved by proving that, with respect to positive infinitesimals x and y

$$(3.2) \qquad D_\varphi(x,y) = o(1) .$$

We shall find it convenient to abbreviate the notation for an integral by writing

$$(3.3) \qquad \int_a^{a'}\int_b^{b'} g(u,v)\, du\, dv = \int\begin{bmatrix}a', & b'\\ a, & b\end{bmatrix}(g) .$$

One of the lemmas essential to showing that the conditions L_R are sufficient for the convergence of S_{mn} to s is Lemma 3.3. The proof of the uniform condition (3.13)" in this lemma may be made to depend upon an application of the Heine-Borel covering theorem to a function J_k of an interval [b,b'], a variable x and a parameter k with values $J_k{}_b^{b'}(x)$. Such a function will appear, for example, in one of the forms

$$(3.4) \qquad \left|\int\begin{bmatrix}x, & b'\\ 0, & b\end{bmatrix}(\varphi)\right| \qquad\qquad P\begin{bmatrix}\pi - x, & b'\\ kx, & b\end{bmatrix}(\varphi)$$

$$P\begin{bmatrix} x, & b' \\ 0, & b \end{bmatrix}(\overline{\varphi}) \qquad P\begin{bmatrix} \pi - x, & b' \\ kx, & b \end{bmatrix}(\overline{\Delta_x g}) .$$

<u>We suppose</u> $J_k{}_b^{b'}(x)$ <u>defined for</u> $[b, b']$ <u>a subinterval of</u> $[0, \pi]$, <u>for</u> $1 < k < \infty$, <u>and</u> $0 < kx < \delta$ <u>for some</u> δ. <u>As a matter of convenience we define</u> $J_k{}_b^{b'}(x)$ <u>as</u> $J_k{}_{b'}^{b}(x)$ <u>and assume that</u>

$$0 \leqq J_k{}_b^{b'}(x) \leqq J_k{}_b^{c}(x) + J_k{}_c^{b'}(x) \tag{3.5}$$

<u>for</u> b, b' <u>and</u> c <u>in</u> $[0, \pi]$ <u>and that for fixed</u> b, b', x, $J_k{}_b^{b'}(x)$ <u>decreases monotonically as</u> k <u>increases</u>.

Lemma 3.1 will be applied to functions of type (3.4) and may be applied to the corresponding functions found in Gergen (1) with V replacing P. In Lemma 3.1 we refer to a function f with numerical values $f(x) > 0$, defined for x small and positive.

LEMMA 3.1. For principal variables $x > 0$, $y > 0$ and for $f(x) > 0$ suppose that

$$J_k{}_b^{b'}(x) = \bar{O}[f(x)] \qquad [(b' - b) = y] \; [0 < b < b' \leqq \pi] \tag{3.6}$$

when either b or b' is fixed and $y = b' - b$ variable, and that

$$J_k{}_0^{y}(x) = \bar{O}[y\, f(x)] \qquad [y > 0] . \tag{3.7}$$

There then exist positive constants x_0, k_0, A such that

$$J_k{}_0^{y}(x) \leqq Ay\, f(x) \tag{3.8}$$

for $0 < x < x_0$, $0 < y \leqq \pi$, $k > k_0$

Given $c \in [0, \pi]$ there exist positive constants k_c, δ_c, A_c and a subinterval I(c) of $[0, \pi]$ open relative to $[0, \pi]$, containing c, and such that for $k > k_c$, $0 < x < \delta_c$, and β, β' in I(c) with $\beta < \beta'$

$$J_k{}_\beta^{\beta'}(x) \leqq A_c\, f(x) . \tag{3.9}$$

This follows from (3.6) when $c > 0$, and from (3.7) when $c = 0$, taking account of (3.5).

By virtue of the Heine-Borel principle there exists a finite set of values $0 = c_0 < c_1 < c_2 < \dots < c_n = \pi$ such that Union $I(c_i) = [0, \pi]$ while each $I(c_i)$ intersects its predecessor for $i = 1, \dots, n$. Given an arbitrary $y \in [0, \pi]$ there exists a chain of points $y_0 < y_1 < \dots < y_r$ such that $0 = y_0$ and $y = y_r$ while both y_i and y_{i-1} are in $I(c_{i-1})$ for $i = 1, \dots, r$. Set

$$(3.10) \qquad k_0 = \max k_{c_i}, \qquad \delta_0 = \min \delta_{c_i}, \qquad A_0 = A_{c_0} + A_{c_1} + \quad + A_{c_n} .$$

Then for $k > k_0$, $0 < x < \delta_0$

$$(3.11) \qquad J_k {}_0^y(x) \leqq \sum_{i=1}^{r} J_k {}_{y_{i-1}}^{y_i}(x) \leqq A_0 f(x)$$

by virtue of (3.10). Combining (3.11) appropriately with (3.7) relation (3.8) follows.

PROOF of Theorem 3.1. We begin with a lemma proved by Gergen (1) p. 39.

LEMMA 3.2. If (C_1) holds and φ is in $L(I)$ then for $a > 0 \quad b > 0$.

$$(3.12)' \qquad \int \begin{bmatrix} ax , by \\ 0 \ , \ 0 \end{bmatrix} (\varphi K^{x,y}) = o(1)$$

$$(3.12)'' \qquad \int \begin{bmatrix} \pi \ , \ \pi \\ \pi - ax , \pi - by \end{bmatrix} (\varphi K^{x,y}) = o(1)$$

Integration by parts suffices to prove (3.12)'. Formula (3.12)" is a consequence of the boundedness of $K^{x,y}(u,v)$ for arguments u,v bounded from zero, and of the integrability of φ.

Lemma 3.3 parallels Lemma 2.2 of p. 40 of Gergen (1) but differs in its statement and proof.

LEMMA 3.3. If (C_1) and L_R'' hold then

$$(3.13)' \qquad \left| \int \begin{bmatrix} x , b' \\ 0 , b \end{bmatrix} (\varphi) \right| = o(x) \qquad [0 < b < b' \leqq \pi]$$

for fixed b or b' and variable $y = b' - b$. Moreover

$$(3.13)'' \qquad \left| \int \begin{bmatrix} x , y \\ 0 , 0 \end{bmatrix} (\varphi) \right| \leqq A\, xy \qquad (0 < x \leqq \pi)\ (0 < y \leqq \pi)$$

for some constant A.

We rewrite (9.6) of MT (11) in the form

$$(3.14)\quad \left|\int\begin{bmatrix} a',b' \\ a\ ,b \end{bmatrix}(g)\right| \leq P\begin{bmatrix} a',b \\ a\ ,b_o \end{bmatrix}\left\{\overline{\Delta_y g}\right\} + \left|\int\begin{bmatrix} a',b_o+y \\ a\ ,b_o \end{bmatrix}(g)\right| \quad [b'-b = y]$$

interchanging the role of x and y, [a,a'] and [b,b']. In (3.14) we now set $a = 0$, $a' = x$, $b_o = ky$, $g = \varphi$. Relation (3.14) then becomes

$$(3.15)\quad \left|\int\begin{bmatrix} x\ ,\ b' \\ 0\ ;\ b \end{bmatrix}(\varphi)\right| \leq P\begin{bmatrix} x\ ,\ b'-y \\ 0\ ,\ ky \end{bmatrix}\left\{\overline{\Delta_y \varphi}\right\} + \left|\int\begin{bmatrix} x,\ ky+y \\ 0,\ ky \end{bmatrix}(\varphi)\right|$$

$$(3.16)\quad \leq \pi P\begin{bmatrix} x\ ,\pi-y \\ 0\ ,\ ky \end{bmatrix}\left\{\overline{\tfrac{1}{v}\Delta_y \varphi}\right\} + \left|\int\begin{bmatrix} x,\ ky+y \\ 0,\ ky \end{bmatrix}(\varphi)\right|$$

where the introduction of the internal factor $\frac{1}{v}$ is compensated for by the external factor π in accordance with Th. 6.2 MT (11). The terms on the right of (3.15) are $\bar{o}(x)$ and $\bar{o}(x\ y)$ by virtue of L''_R and (C_1) respectively. Since the left member of (3.16) is independent of k it becomes $o(x)$ and (3.13)' follows.

To obtain (3.13)" Lemma 3.1 will be used, setting $J\,_b^{b'}(x)$ equal to the left member of (3.13)'. With $f(x) = x$, (3.6) is satisfied by virtue of (3.13)', and (3.7) is satisfied on account of (C_1). According to (3.8) then

$$(3.17)\quad \left|\int\begin{bmatrix} x,y \\ 0,0 \end{bmatrix}(\varphi)\right| \leq A\ xy$$

for suitable A and $0 < x < x_o$, $0 < y \leq \pi$.

One can similarly establish (3.13)' with the roles of x and y interchanged, and it will follow that (3.17) holds for suitable A and $0 < y < y_o$, $0 < x \leq \pi$. It is clear that (3.13)" then holds as stated.

Modifying Gergen's notation slightly set

$$(3.18)'\quad \omega^y(v) = 2\csc\left(\frac{v+y}{2}\right) - \csc\left(\frac{v+2y}{2}\right) - \csc\left(\frac{v}{2}\right)$$

$$(3.18)''\quad \Omega^{x,y}(u,v) = \sin\left(\frac{\pi\ u}{x}\right)\ \csc\left(\frac{u}{2}\right)\ \sin\left(\frac{\pi\ v}{y}\right)\ \omega^y(v)$$

$$(3.18)'''\quad W^{x,y}(u,v) = \omega^x(u)\ \sin\left(\frac{\pi\ u}{x}\right)\ \omega^y(v)\ \sin\left(\frac{\pi\ v}{y}\right).$$

We shall refer to the relation

$$(3.19) \qquad v^3\,|\omega^y(v)| < A\,y^2 \qquad\qquad v^4\,|\omega^y_v(v)| < A\,y^2$$

readily established for $0 < y < v \leq \pi - y$. See Gergen p. 42. By a straightforward integration by parts Gergen shows [(1) pp. 41, 44, 45] that *whenever* (3.13)" holds

$$(3.20) \qquad \int\begin{bmatrix} ax , & \pi \\ 0 \;, & \pi - by \end{bmatrix} (\varphi K^{x,y}) = o(1) \qquad [a > 0,\ b > 0]$$

$$(3.21) \qquad \int\begin{bmatrix} ax , & \pi - 2y \\ 0 \;, & ky \end{bmatrix} (\varphi \Omega^{x,y}) = \bar{o}(1)$$

$$(3.22) \qquad \int\begin{bmatrix} \pi - 2x , & \pi \\ kx \;, & \pi - by \end{bmatrix} (\varphi \Omega_1^{x,y}) = \bar{o}(1) \qquad [\Omega_1^{x,y}(u,v) = \Omega^{y,x}(v,u)]$$

$$(3.23) \qquad \int\begin{bmatrix} \pi - 2x , & \pi - 2y \\ kx \;, & ky \end{bmatrix} w^{x,y}\,\psi^{x,y} = \bar{o}(1)$$

where $\psi^{x,y}(u,v) = \varphi(u+x\,,\,v+y)$. The symmetry of the conditions (3.13)" imply that relations[6] similar to (3.20) to (3.23) hold with x and y interchanged when (3.13)" holds.

Lemma 3.4 parallels Gergen's Lemma 5, Gergen (1), p. 42. Its proof however depends in an essential way upon the new lemmas on Fréchet variation of MT (11).

LEMMA 3.4. If (C_1) and L_R'' hold then

$$(3.24) \qquad J = \int\begin{bmatrix} ax , & \pi \\ 0 \;, & 0 \end{bmatrix} (\varphi K^{x,y}) = o(1) \qquad (a > 0).$$

Let $Q = \begin{bmatrix} ax , & \pi - 2y \\ 0 \;, & ky \end{bmatrix}$ and let Q_y and Q_{2y} be the y- and 2y- translations of Q namely

$$Q_y = \begin{bmatrix} ax , & \pi - y \\ 0 \;, & (k+1)y \end{bmatrix} \;; \qquad Q_{2y} = \begin{bmatrix} ax , & \pi \\ 0 \;, & (k+2)y \end{bmatrix} .$$

It follows from (3.20) and Lemma 3.2 [Cf. Gergen] that J differs by $\bar{o}(1)$ from the integral of $\varphi K^{x,y}$ over Q, Q_y and Q_{2y}, so that

$$(3.25) \qquad 4J = \int[Q, \varphi K^{x,y}] + 2\int[Q_y, \varphi K^{x,y}] + \int[Q_{2y}, \varphi K^{x,y}] + \bar{o}(1) .$$

---

6. The relations obtained from (3.20) to (3.23) by interchanging x and y will be called *conjugate* to (3.20) and (3.23). If (X) designates a given relation (X)* will be used to designate the conjugate relation.

Changes of the variable v of integration by virtue of which Q_{2y} and Q_y have the image Q will lead to the relation

$$(3.26)\qquad 4J = H^{x,y} - \int\begin{bmatrix}ax, & \pi-2y\\ 0, & ky\end{bmatrix}(\varphi\,\Omega^{x,y}) + \bar{o}(1) = H^{x,y} + \bar{o}(1)$$

by (3.25) and (3.21) where

$$H^{x,y} = \int_0^{ax} \frac{\sin\left(\frac{\pi u}{x}\right)du}{\sin\left(\frac{u}{2}\right)} \int_{ky}^{\pi-2y}\left[\frac{\Delta_{2y}[\varphi(u,v)]}{\sin\left(\frac{v+2y}{2}\right)} - 2\,\frac{\Delta_y[\varphi(u,v)]}{\sin\left(\frac{v+y}{2}\right)}\right]\sin\left(\frac{\pi v}{y}\right)dv.$$

Let $F^{x,y}(u,v)$ be the integrand of $H^{x,y}$. According to (9.5) MT (11)

$$(3.27)\qquad |H^{x,y}| \leq P\begin{bmatrix}ax, & \pi-2y\\ 0, & ky\end{bmatrix}(\overline{F^{x,y}})$$

We shall use Th. 6.2 MT (11) to simplify the right member of (3.27). With the variables u, v, x, y limited as in the integral $H^{x,y}$

$$(3.27)'\qquad \frac{\sin\left(\frac{\pi u}{x}\right)}{\sin\left(\frac{u}{2}\right)} \leq \frac{\left(\frac{\pi u}{x}\right)}{\sin\left(\frac{u}{2}\right)} \leq \frac{\pi^2}{x} \qquad (x > 0,\ u > 0)$$

$$(3.27)''\qquad \frac{\sin\left(\frac{\pi v}{y}\right)}{\sin\left(\frac{v+y}{2}\right)} \leq \frac{1}{\sin\left(\frac{v}{2}\right)} \leq \frac{\pi}{v} \qquad (y > 0,\ v > 0)$$

It follows from Th. 6.2 MT (11) that

$$|H^{x,y}| \leq \frac{\pi^3}{x}\left[P\begin{bmatrix}ax, & \pi-2y\\ 0, & ky\end{bmatrix}\left(\overline{\tfrac{1}{v}\Delta_{2y}\varphi}\right) + 2P\begin{bmatrix}ax, & \pi-2y\\ 0, & ky\end{bmatrix}\left(\overline{\tfrac{1}{v}\Delta_{y}\varphi}\right)\right]$$

Hence $H^{x,y} = \bar{o}(x)/x = \bar{o}(1)$ by virtue of condition L_R''. Hence $J = \bar{o}(1) = o(1)$ since the left member of (3.24) is independent of k.

Lemma 3.5 replaces Lemma 6 of Gergen (1) p. 43.

LEMMA 3.5. If L_R' and L_R'' hold then

$$(3.28)\qquad P\begin{bmatrix}\pi-x, & b'\\ kx, & b\end{bmatrix}\left\{\overline{\tfrac{1}{u}\Delta_x\varphi}\right\} = \bar{o}(1) \qquad (0 < b < b' \leq \pi)$$

for fixed b or b' and variable $b' - b = y$. Moreover positive constants x_0, k_0, A exist such that

$$P\begin{bmatrix}\pi-x, & y\\ kx & , 0\end{bmatrix}\left\{\overline{\frac{1}{u}\Delta_x\varphi}\right\} \leq A\,y \tag{3.29}$$

for $0 < x < x_0$, $k_0 < k$, $0 < y \leq \pi$.

In (9.12) of MT (11) set $\omega = \frac{1}{u}\Delta_x\varphi$ and $b_0 = ky$, $a = kx$, $a' = \pi - x$. With $b' - b = y$ and b' or b fixed

$$P\begin{bmatrix}\pi-x, & b'\\ kx & , b\end{bmatrix}\left\{\overline{\frac{1}{u}\Delta_x\varphi}\right\} \leq P\begin{bmatrix}\pi-x, & b'-y\\ kx & , ky\end{bmatrix}\left\{\overline{\frac{1}{u}\Delta_{xy}\varphi}\right\} + P\begin{bmatrix}\pi-x, & (k+1)y\\ kx & , ky\end{bmatrix}\left\{\overline{\frac{1}{u}\Delta_x\varphi}\right\}$$

which is at most $\bar{o}(1)$ by L_R' and L_R''. This establishes (3.28).

We apply Lemma 3.1 to obtain (3.29). One sets $J_k{}_b^{b'}(x)$ equal to the left member of (3.28). It follows from (3.28) that (3.6) is satisfied with $f(x) \equiv 1$, while L_R'' then implies (3.7). Relation (3.29) follows from (3.8).

Lemma 3.6 corresponds to Lemma 7 p. 44 of Gergen (1).

LEMMA 3.6. If L_R holds then

$$\int\!\!\int\begin{bmatrix}\pi & ,\pi\\ 0 & ,\pi-by\end{bmatrix}(\varphi K^{x,y}) = o(1) \qquad (b > 0). \tag{3.30}$$

As in the proof of Lemma 3.4 one finds that

$$4\int\!\!\int\begin{bmatrix}\pi & ,\pi\\ 0 & ,\pi-by\end{bmatrix}(\varphi K^{x,y}) = H_1^{x,y} - \int\!\!\int\begin{bmatrix}\pi-2x, & \pi\\ kx & ,\pi-by\end{bmatrix}(\varphi\Omega_1^{x,y}) + \bar{o}(1) = H_1^{x,y} + \bar{o}(1 \tag{3.31}$$

using (3.22), where

$$H_1^{x,y} = \int_{\pi-by}^{\pi}\frac{\sin\left(\frac{\pi v}{y}\right)}{\sin\left(\frac{v}{2}\right)}dv\int_{kx}^{\pi-2x}\left[\frac{\Delta_{2x}\varphi(u,v)}{\sin\left(\frac{u+2x}{2}\right)} - \frac{2\Delta_x\varphi(u,v)}{\sin\left(\frac{u+x}{2}\right)}\right]\sin\left(\frac{\pi u}{x}\right)du$$

Relations similar to (3.27)' and (3.27)" hold here with u and x replacing v and y. It then follows from (9.5) and Th. 6.2 of MT (11) that with $\rho = \sin\frac{\pi-by}{2}$

$$|H_1^{x,y}| \leq \frac{\pi}{\rho}\left[P\begin{bmatrix}\pi-2x, & \pi\\ kx & ,\pi-by\end{bmatrix}\left\{\overline{\frac{1}{u}\Delta_{2x}\varphi}\right\} + 2P\begin{bmatrix}\pi-2x, & \pi\\ kx & ,\pi-by\end{bmatrix}\left\{\overline{\frac{1}{u}\Delta_x\varphi}\right\}\right]. \tag{3.32}$$

Since $\pi - by$ may be considered bounded from zero the right member of (3.32) is $\bar{o}(1)$ by virtue of (3.28). Hence (3.30) holds with a right member $\bar{o}(1)$. But the left member of (3.30) is independent of k so that (3.30) holds as stated.

A translation principle. We shall state a principal already implicitly used and to be used systematically in the following lemmas and theorems. Let T_y denote the translation of the points on the v-axis which replaces the point v by v+y. If V is an interval of the v-axis T_yV shall denote the translated interval. If $Q = U \times V$ is a 2-dimensional interval T_yQ shall denote the interval $U \times T_yV$. If k is defined over Q with values k(u,v), and k(u, v+y) is also defined over Q, we denote the function with values k(u, v+y) by T_yk. The inverse translation T_y^{-1} is similarly used, replacing y by -y. With this understood the translation principle states that

$$(3.33) \qquad P(Q,k) = P(T_y^{-1} Q, T_yk)$$

and follows directly from the definition of the F-variation. In applying this principle we shall use the fact that $T_y(\bar{k})$ and $\overline{T_yk}$, differ by functions of u and v alone and so have the same F-variation over a common interval.

The translation principle will be used in proving Lemma 3.7. Cf. Lemma 9, Gergen (1) p. 45.

LEMMA 3.7. If L_R' and L_R'' hold, and if $F^{x,y}$ has the values

$$(3.34) \qquad F^{x,y}(u,v) = \frac{\omega^y(v)}{u} \Delta_x \varphi(u, v+y) \qquad \left\{(u,v) \in \begin{bmatrix} \pi, & \pi - y \\ 0+, & 0 \end{bmatrix}\right\}$$

then

$$(3.35) \qquad J_1 = P \begin{bmatrix} \pi - x, & \pi - 2y \\ kx & ,ky \end{bmatrix} \left\{\overline{F^{x,y}}\right\} = \bar{o}(1) .$$

By the translation principle

$$(3.36) \qquad J_1 = P \begin{bmatrix} \pi - x, & \pi - y \\ kx & ,(k+1)y \end{bmatrix} \left\{\overline{f^{x,y}}\right\} \quad \left[\text{where } f^{x,y}(u,v) = \frac{\omega^y(v-y)}{u} \Delta_x \varphi(u,v)\right] .$$

In accordance with (6.20) of MT (11), $J_1 \leq J' + J''$ where

$$(3.37) \qquad J' = |\omega^y(\pi - 2y)| \; P \begin{bmatrix} \pi - x, & \pi - y \\ kx & ,(k+1)y \end{bmatrix} \left\{\overline{\frac{1}{u} \Delta_x \varphi}\right\}$$

$$J'' = \int_{(k+1)y}^{\pi - y} |\omega_v^y(v-y)| \; P \begin{bmatrix} \pi - x, & v \\ kx & ,(k+1)y \end{bmatrix} \left\{\overline{\frac{1}{u} \Delta_x \varphi}\right\} dv .$$

Since $\omega^y(\pi - 2y) = O(y^2)$ by (3.19), and since (3.29) holds, we infer that

$J' = \bar{O}(y^2)$. In the case of J" use is made of the relation $v^4|\omega_v^y(v)| < A\, y^2$ given by (3.19). As a consequence

$$|\omega_v^y(v-y)| < A\left[\frac{y^2}{(v-y)^4}\right]$$

and using (3.29)

$$J'' = \bar{O}\left[y^2 \int_{(k+1)y}^{\pi-y} \frac{v\,dv}{(v-y)^4}\right]$$

$$= \bar{O}\left[y^2 \int_{ky}^{\pi} \frac{(v+y)dv}{v^4}\right] = \bar{o}(1)$$

Thus (3.35) holds as stated.

We shall refer to the relation obtained from (3.35) by an application of the translation principle.

(3.38) $J_2 = P\begin{bmatrix}\pi-2x, \pi-2y\\ kx \quad ,ky\end{bmatrix}\{\overline{F_{xy}}\} = \bar{o}(1)$ [where $F_{xy}(u,v) = \frac{\omega^y(v)}{u+x}\Delta_x\,\varphi(u+x,v+y)$]

<u>Summary of preceding lemmas</u>. Lemmas 3.2, 3.4, 3.6 and relation (3.20) concern integrals with the integrand of the Dirichlet integral $D_\varphi(x,y)$, taken over rectangles adjacent to an edge or a vertex of $\begin{bmatrix}\pi,\pi\\ 0,0\end{bmatrix} = I$. According to these lemmas for $a > 0$, $b > 0$

(3.39) $$\int\begin{bmatrix}ax, by\\ 0 \ , \ 0\end{bmatrix}(\varphi K^{x,y}) = o(1)$$

(3.40) $$\int\begin{bmatrix}\pi \quad\ , \pi\\ \pi-ax, \pi-by\end{bmatrix}(\varphi K^{x,y}) = o(1)$$

(3.41) $$\int\begin{bmatrix}ax, \pi\\ 0 \ , \ 0\end{bmatrix}(\varphi K^{x,y}) = o(1)$$

(3.42) $$\int\begin{bmatrix}\pi, \pi\\ 0, \pi-by\end{bmatrix}(\varphi K^{x,y}) = o(1)$$

(3.43) $$\int\begin{bmatrix}ax, \pi\\ 0 \ , \pi-by\end{bmatrix}(\varphi K^{x,y}) = o(1) .$$

Corresponding to the last three relations there are three other relations obtained by interchanging the roles of x and y in the limits. In all there are four corner rectangles and four rectangles adjacent to the four edges of I. We shall use these relations in proving the principal theorem of this section.

THEOREM 3.1. When $\varphi = f - s$ satisfies L_R over I, then the Dirichlet integral $D_\varphi(x,y) = o(1)$.

We begin with the formulas used in Gergen's proof of his Th. 1. It follows from the eight relations in the above summary that

$$(3.44) \qquad 16\, D_\varphi(x,y) = \left\{\int_{kx}^{\pi-2x} + 2\int_{(k+1)x}^{\pi-x} + \int_{(k+2)x}^{\pi}\right\} Q_k^{x,y}(u)\, du + \bar{o}(1)$$

where

$$(3.45) \qquad Q_k^{x,y}(u) = \left\{\int_{ky}^{\pi-2y} + 2\int_{(k+1)y}^{\pi-y} + \int_{(k+2)y}^{\pi}\right\} \varphi(u,v)\, K^{x,y}(u,v)\, dv\,.$$

On making changes of variables of integration carrying the respective rectangles of integration into $\begin{bmatrix}\pi-2x, & \pi-2y\\ kx & ,ky\end{bmatrix}$ we have

$$(3.46) \qquad 16\, D\varphi(x,y) = \int_{kx}^{\pi-2x} \sin(\pi u/x)\, du \int_{ky}^{\pi-2y} \psi^{x,y}(u,v) \sin(\pi v/y)\, dv + \bar{o}(1)$$

where

$$(3.47) \qquad \psi^{x,y}(u,v) = S_1^{x,y}(u,v) + S_2^{x,y}(u,v) + S_3^{x,y}(u,v) + S_4^{x,y}(u,v)$$

with

$$S_1^{x,y}(u,v) = \frac{\Delta_{xy}\varphi(u+x,\, v+y)}{\sin\left(\frac{u+2x}{2}\right)\sin\left(\frac{v+2y}{2}\right)} - \frac{\Delta_{xy}\varphi(u\,,\, v+y)}{\sin\left(\frac{u}{2}\right)\sin\left(\frac{v+2y}{2}\right)}$$

$$\frac{\Delta_{xy}\varphi(u+x\,,\, v)}{\sin\left(\frac{u+2x}{2}\right)\sin\left(\frac{v}{2}\right)} + \frac{\Delta_{xy}\varphi(u,\, v)}{\sin\left(\frac{u}{2}\right)\sin\left(\frac{v}{2}\right)}$$

$$S_2^{x,y}(u,v) = -\omega^y(v)\left\{\frac{\Delta_x\varphi(u+x,v+y)}{\sin\left(\frac{u+2x}{2}\right)} - \frac{\Delta_x\varphi(u,v+y)}{\sin\left(\frac{u}{2}\right)}\right\}$$

$$S_3^{x,y}(u,v) = -\omega^x(u)\left\{\frac{\Delta_y\varphi(u+x\,,\, v+y)}{\sin\left(\frac{v+2y}{2}\right)} - \frac{\Delta_y\varphi(u+x,\, v)}{\sin\left(\frac{v}{2}\right)}\right\}$$

$$S_4^{x,y}(u,v) = \omega^x(u)\ \omega^y(v)\ \varphi(u+x,\ v+y).$$

The contribution of $S_1^{x,y}$ to $D_\varphi(x,y)$ has the form

$$(3.48) \qquad D_1^{x,y} = \int_{kx}^{\pi-2x} \sin(\pi u/x)du \int_{ky}^{\pi-2y} S_1^{x,y}(u,v) \sin(\pi v/y)\ dv\ .$$

According to (9.5) MT (11), and with the aid of Th. 6.2 MT (11)

$$|D_1^{x,y}| \leq P\begin{bmatrix}\pi-2x\ ,\pi-2y\\ kx \quad ,ky\end{bmatrix} \ (\overline{S_1^{x,y}})$$

since $|\sin \pi u/x| \leq 1$, $|\sin \pi v/y| \leq 1$.

After an x and y translation of the two variables of integration, the contribution of the first term in $S_1^{x,y}$ to $D_1^{x,y}$ reduces to

$$(3.49) \qquad \bar{o}\left[P\begin{bmatrix}\pi-x\ ,\ \pi-y\\ (k+1)x,\ (k+1)y\end{bmatrix}\left\{\overline{\frac{1}{uv}\Delta_{xy}\varphi}\right\}\right].$$

In obtaining this result Th. 6.2 of MT (11) has been used to justify replacing $\sin\left(\frac{u+2x}{2}\right)$ by u in the denominator of $S_1^{x,y}(u,v)$, noting that for (u,v) in the interval of (3.49)

$$\frac{1}{\sin\left(\frac{u+2x}{2}\right)} \leq \frac{1}{\sin\ \left(\frac{u}{2}\right)} \leq \frac{\pi}{u}$$

The replacing of $\sin\left(\frac{v+2y}{2}\right)$ by v is similarly justified. It follows from L_R' that the contribution of (3.49) to $D_1^{x,y}$ is $\bar{o}(1)$. The other terms in $S_1^{x,y}$ make similar contributions to $D_\varphi(x,y)$. The contribution of the first term in $S_2^{x,y}$ to $D_2^{x,y}$ can be simplified by replacing $\sin \frac{1}{2}(u+2x)$ by u+x using Th. 6.2 MT (11). So reduced this term equals $O(J_2) = \bar{o}(1)$ where J_2 is given by (3.38). The remaining term in $S_2^{x,y}$ similarly reduces to $O(J_1) = \bar{o}(1)$ where J_1 is given by (3.35). The contribution of $S_3^{x,y}$ to $D_3^{x,y}$ is likewise $\bar{o}(1)$. Finally $D_4^{x,y} = \bar{o}(1)$ by (3.23).

Thus $D_i^{x,y} = \bar{o}(1)$ for i = 1, ..., 4. Hence $D_\varphi(x,y) = o(1)$ and the proof of the theorem is complete.

4. <u>Proof that</u> $Y' \Rightarrow L_P' + L_P''$. We are assuming $f \in L$ over $\begin{bmatrix}\pi, \pi\\ 0, 0\end{bmatrix}$. The condition Y' requires that the function with values

$$\theta(u,v) = uv\ \varphi(u,v) \qquad \left[(u,v) \in \begin{bmatrix} \pi, & \pi \\ 0, & 0 \end{bmatrix}\right] \tag{4.1}$$

satisfy the condition

$$P\begin{bmatrix} x, & y \\ 0, & 0 \end{bmatrix}(\theta) \leq A\, xy \qquad \left[(x,y) \in \begin{bmatrix} \pi, & \pi \\ 0+, & 0+ \end{bmatrix}\right]. \tag{4.2}$$

We note the following.

LEMMA 4.1. The function θ is continuous at each point $(u,v) \in \begin{bmatrix} \pi, & \pi \\ 0, & 0 \end{bmatrix}$ at which $u = 0$, or $v = 0$.

The mixed difference

$$|\theta(x,y) - \theta(x,0) - \theta(0,y) + \theta(0,0)| = |\theta(x,y)| \tag{4.3}$$

since $\theta(u,v)$ vanishes on the axes. But by definition of the F-variation, and by (4.2)

$$|\theta(x,y)| \leq P\begin{bmatrix} x, & y \\ 0, & 0 \end{bmatrix}(\theta) \leq A\,x\,y \tag{4.4}$$

from which the lemma follows.

We begin the proof that $Y_p \Rightarrow L_p'$ with a formal lemma.

LEMMA 4.2. If g satisfies $\hat{F}$ over $Q = \begin{pmatrix} a', & b' \\ a, & b \end{pmatrix}$ and if $a < a'-x < a'$, $b < b'-y < b'$, then

$$J = P\begin{bmatrix} a'-x, & b'-y \\ a, & b \end{bmatrix}(\overline{\Delta_{xy}\, g}) \leq 36\ xy\ P(Q,\, g)\ . \tag{4.5}$$

We shall make use of a special evaluation of an F-variation of an integral. Let $C[c,c']$ represent the Banach space of functions continuous over $[c,c']$, and let elements in $C[c,c']$ whose norm is at most 1 be termed <u>subunit</u>. According to Lemma 6.1 of MT (11)

$$J = \sup_{\lambda,\mu} \int_a^{a'-x} \int_b^{b'-y} \lambda(u)\,\mu(v)\,\Delta_{xy}\big(g(u,v)\big)\,du\,dv = \sup_{\lambda,\mu} H \tag{4.6}$$

taking the sup over all subunit $\lambda \in C[a,a']$ and subunit $\mu \in C[b,b']$. It is convenient to use the spaces $C[a,a']$ and $C[b,b']$ rather than the spaces $C[a,a'-x]$, $C[b,b'-y]$. Cf. proof of Lemma 4.3. It is clear that

$$H = \int_a^{a'-x} \lambda(u)\, du \int_b^{b'-y} \mu(v)\, [g(u+x,v+y)-g(u+x,v)-g(u,v+y)+g(u,v)]\, dv .$$

On making the appropriate changes in the variables of integration H takes the form

$$\left[\int_{a+x}^{a'} \lambda(u-x) - \int_a^{a'-x} \lambda(u)\right] \left[\int_{b+y}^{b'} \mu(v-y) - \int_b^{b'-y} \mu(v)\right] g(u,v)\, dudv =$$

$$(4.7) \qquad \left[\int_{a'-x}^{a'} \lambda(u-x) + \int_{a+x}^{a'-x} [\lambda(u-x)-\lambda(u)] - \int_a^{a+x} \lambda(u)\right]$$

$$\left[\int_{b'-y}^{b'} \mu(v-y) + \int_{b+y}^{b'-y} [\mu(v-y)-\mu(v)] - \int_b^{b+y} \mu(v)\right] g(u,v)\, dudv .$$

This expands into the sum of nine integrals of the form

$$(4.8) \qquad I_i = \int_{a_i}^{a_i'} \int_{b_i}^{b_i'} h_i(u)\, k_i(v)\, g(u,v)\, dudv \qquad (i = 1, \ldots, 9)$$

where the interval Q_i in I_i is a subinterval of Q. For example one may start with

$$I_1 = \int_{a'-x}^{a'} \int_{b'-y}^{b'} \lambda(u-x)\, \mu(v-y)\, g(u,v)\, dudv .$$

We shall condition the respective integrals I_i by means of an inequality introduced by the authors as a replacement for the second law of the mean for double integrals. As a matter of notation when h is in L over $[c,c']$ we have set

$$\max_{\alpha} \left| \int_{\alpha}^{c'} h(u)\, du \right| = M_c^{c'}(h) \qquad \{\alpha \in [c,c']\} .$$

We now extend the definitions of h_i and k_i over the intervals $[a,a']$ and $[b,b']$ respectively, by setting $h_i(u) = 0$ for $u \in [a,a'] - [a_i,a_i']$ and $k_i(v) = 0$ for $v \in [b,b'] - [b_i,b_i']$. With this definition we note that $M_{a_i}^{a_i'}(h_i) = M_a^{a'}(h_i)$ and $M_{b_i}^{b_i'}(k_i) = M_b^{b'}(k_i)$. Lemma 5.2 MT (11) applies to Q if $g(a+, t) = 0$ for $t \in (b,b')$ and $g(s,b+) = 0$ for $s \in (a,a')$. But in proving (4.5) no generality is lost if one assumes that g has this property; for the function with values

$$g'(u,v) = g(u,v) - g(a+,v) - g(u,b+) + g(a+,b+) \qquad [(u,v) \in Q] \tag{4.9}$$

has this property and substituted for g in (4.5) alters neither side of (4.5). Assuming accordingly that g satisfies the hypotheses of Lemma 5.2 MT (11), applied to Q,

$$I_i \leq M_{a_i}^{a_i'}(h_i)\, M_{b_i}^{b_i'}(k_i)\, P(Q, g) \tag{4.10}$$

$$\leq 4\, xy\, P(Q, g) \tag{4.11}$$

as we shall see. To obtain (4.11) from (4.10) we have merely to verify the relations ($i = 1, \ldots, 9$)

$$(4.12)' \quad M_{a_i}^{a_i'}(h_i) \leq 2x \; ; \qquad (4.12)'' \quad M_{b_i}^{b_i'}(k_i) \leq 2y \; .$$

The relations (4.12) are obvious whenever $a_i' - a_i = x$ or $b_i' - b_i = y$, since λ and μ are subunit elements in their Banach spaces. All other elements in (4.12)' have limits $a_i = a+x$, $a_i' = a' - x$ and so are the max for $\alpha \in [a+x, a'-x]$ of

$$\left| \int_{\alpha}^{a'-x} (\lambda(u-x) - \lambda(u))\, du \right| = \left| \int_{\alpha-x}^{a'-2x} \lambda(u)\, du - \int_{\alpha}^{a'-x} \lambda(u)\, du \right|$$

$$= \left| \int_{\alpha-x}^{\alpha} \lambda(u)\, du - \int_{a'-2x}^{a'-x} \lambda(u)\, du \right| \leq 2x$$

A similar proof satisfies (4.12)", and (4.5) follows from (4.11).

COROLLARY 4.0. If g is in L over $Q = \begin{pmatrix} a', & b' \\ a & ; & b \end{pmatrix}$ and $a < a'-x < a'$, $b < b'-y < b'$ then (4.5) holds.

In case $P(Q, g) = \infty$ the corollary is trivial. Suppose then that $P(Q, g) < \infty$. If g satisfies $\hat{F}$ over Q the corollary follows from Lemma 4.2. If g does not satisfy $\hat{F}$ over Q then for $a < \alpha < a'$, $b < \beta < b'$, the function g" with values

$$g''(u, v) = g(u, v) - g(\alpha, v) - g(u, \beta) + g(\alpha, \beta)$$

will satisfy $\hat{F}$ over Q. Substituted for g in (4.5), g" will yield the same values of the members of (4.5) as does g. But (4.5) holds for g" by Lemma 4.2, and hence holds for g.

In showing that $Y_P \to L''_P$ a second formal lemma will be used.

LEMMA 4.3. If g is in L over $Q = \begin{pmatrix} a', & b' \\ a, & b \end{pmatrix}$ and $b < b'-y < b'$ then for $a < u < a'$,

$$(4.13) \qquad J_1 = P\begin{bmatrix} a', & b'-y \\ a, & b \end{bmatrix} \left\{\overline{\Delta_y g}\right\} \leqq (a'-a)\ 6\ y \sup_u T_{b+}^{b'-} [g(u,\cdot)] .$$

If the right member of (4.13) is infinite (4.13) holds trivially. In case the right member of (4.13) is finite the proof of (4.13), similar to the proof of Lemma 4.2, is as follows:

Making use of the functions λ and μ of the proof of Lemma 4.3

$$(4.13)' \qquad J_1 = \sup_{\lambda,\mu} \int_a^{a'} \int_b^{b'-y} \lambda(u)\ \mu(v)\Delta_y g(u,v)\ du dv = \sup_{\lambda,\mu} H_1$$

introducing H_1. One sees that with $u \in [a,a']$

$$|H_1| \leqq (a'-a) \sup_u \left| \left[\int_{b+y}^{b'} \mu(v-y) - \int_b^{b'-y} \mu(v)\right] g(u,v)\ dv \right|$$

$$= (a'-a) \sup_u \left| \left[\int_{b'-y}^{b'} \mu(v-y) + \int_{b+y}^{b'-y} [\mu(v-y) - \mu(v)] - \int_b^{b+y} \mu(v)\right] g(u,v)\ dv \right| .$$

Without loss of generality we can assume that $g(u, b+) = 0$ for $u \in (a,a')$; if g does not have this property the function

$$g'(u,v) = g(u,v) - g(u,b+)$$

does, and substituted for g in (4.13) alters neither member of (4.13). Assuming then that $g(u,b+) = 0$ for $u \in (a,a')$, the second law of the mean and (4.12)" give (as in the proof of 4.11)

$$|H_1| \leqq (a'-a)\ 6y \sup_u T_{b+}^{b'-} g(u,\cdot) \qquad [\text{Cf. } (5.2) \text{ MT } (11)]$$

and (4.13) follows from (4.13)'.

The principal theorem of this section follows.

THEOREM 4.1. The condition $Y' \Rightarrow L_P' + L_P''$.

To show that $Y' \Rightarrow L_P'$ use is made of Cor. 4.0 in the form

$$(4.14) \qquad J = P\begin{bmatrix}\pi - x, & \pi - y\\ kx & , ky\end{bmatrix}\left\{\overline{\Delta_{xy}\left(\frac{\varphi}{uv}\right)}\right\} \leq 36\ xy\ P\begin{pmatrix}\pi, & \pi\\ kx, & ky\end{pmatrix}\left[\frac{\theta}{u^2v^2}\right]$$

recalling that $\theta(u,v) = uv\ \varphi(u,v)$. By hypothesis

$$(4.15) \qquad P\begin{bmatrix}u, & v\\ 0, & 0\end{bmatrix}(\theta) \leq A\ uv \qquad \left\{(u,v) \in \begin{bmatrix}\pi, & \pi\\ 0+, & 0+\end{bmatrix}\right\}$$

and it follows from Cor. 4.3, MT (11) that for fixed positive k with $0 < kx < \pi - x$, $0 < ky < \pi - y$,

$$(4.16) \qquad P\begin{bmatrix}\pi, & \pi\\ kx, & ky\end{bmatrix}\left\{\frac{\theta}{u^2v^2}\right\} = \frac{A}{k^2}\ O\left(\frac{1}{xy}\right).$$

We infer from (4.14) and (4.16) that for fixed k

$$J = \frac{A}{k^2}\ O(1)$$

from which it follows that $J = \bar{o}(1)$.

To show that $Y' \Rightarrow L_P''$, use is made of (4.13) in the form

$$(4.17) \qquad J_1 = P\begin{bmatrix}x, & \pi - y\\ 0, & ky\end{bmatrix}\left[\overline{\Delta_y\ \frac{\varphi}{v}}\right] \leq 6\ xy\ \sup_u T_{ky}^{\pi}\left[\frac{\varphi(u,\cdot)}{v}\right] \qquad (0 < u < x)$$

where $0 < ky < \pi - y$. We see that for fixed k and u

$$T_{ky}^{\pi}\left[\frac{\varphi(u,\cdot)}{v}\right] = T_{ky}^{\pi}\left[\frac{\theta(u,\cdot)}{uv^2}\right] \qquad [u > 0,\ ky > 0].$$

We shall use Cor. 4.1 of MT (11) setting $f(v) = \dfrac{\theta(u,v)}{u}$, $(u > 0)$. Observe that

$$T_0^v[f] = \frac{1}{u}\,T_0^v[\theta(u,\cdot)] \leq \frac{1}{u}\,P\begin{bmatrix}u, & v\\ 0, & 0\end{bmatrix}(\theta) \leq A\ v \qquad (u > 0)$$

so that Cor. 4.1 implies that

$$\frac{A}{k}\, O\left(\frac{1}{y}\right) = T^{\pi}_{ky}\left[\frac{f}{v^2}\right] = T^{\pi}_{ky}\left[\frac{\varphi(u,\cdot)}{v}\right].$$

Combining this result with (4.17) one has $J_1 = \frac{A}{k}\, O(x)$, and we conclude that $J_1 = \bar{o}(x)$.

In a similar manner one shows that the condition in L_P'' conjugate to $J_1 = \bar{o}(x)$ is satisfied. This completes the proof of the theorem.

COROLLARY 4.1. $Y_P \Rightarrow L_P$.

This is clear since $Y_P = Y' + (C_1)$ and $L_P = L_P' + L_P'' + (C_1)$.

5. Proof that $L_P' + L_P'' \Rightarrow C^0$ and $L_2' + L_2'' \Rightarrow C^0$. These relations are instrumental in showing that $L_P \equiv C^0 + L_R$. The conditions $L_P' + L_P''$ are as follows

$$(5.1)' \qquad \zeta(x, y, k) = P\begin{bmatrix} \pi-x, & \pi-y \\ kx & , ky \end{bmatrix}\left\{\overline{\Delta_{xy}\left(\frac{\varphi}{uv}\right)}\right\} = \bar{o}(1)$$

$$(5.1)'' \qquad \xi(x, y, k) = P\begin{bmatrix} x, & \pi-y \\ 0, & ky \end{bmatrix}\left\{\overline{\Delta_{y}\left(\frac{\varphi}{v}\right)}\right\} = \bar{o}(x)$$

$$(5.1)''' \qquad \eta(x, y, k) = P\begin{bmatrix} \pi-x, & y \\ kx & , 0 \end{bmatrix}\left\{\overline{\Delta_{x}\left(\frac{\varphi}{u}\right)}\right\} = \bar{o}(y) .$$

We are accordingly concerned with relations between ζ, ξ, η and the F-variation $P\begin{bmatrix} x, & y \\ 0, & 0 \end{bmatrix}(\hat{\varphi})$ used in defining C^0 and C^0.

The proof of the relation $L_P' + L_P'' \Rightarrow C^0$ will be clear if one distinguishes between the formal laws in which the order relations do not appear, and the relations thereby obtained on imposing the order conditions. We begin with a formal relation [Cf. Gergen (1) Lemma 1, p. 47] involving the variation (5.1)''' where the interval $[0, y]$ is replaced by $[b, b']$. A similar law holds with $[0, x]$ in (5.1)'' replaced by $[a, a']$. The principal formal relation appears in Lemma 5.2.

LEMMA 5.1. With $0 < b < b' \leq \pi$, $b' - b = y$, $0 < kx < \pi - x$, $0 < ky < b$, $k \geq 1$

$$(5.2) \qquad J_3 = P\begin{bmatrix} \pi-x, & b' \\ kx & , b \end{bmatrix}\left\{\overline{\Delta_{x}\left(\frac{\varphi}{u}\right)}\right\} \leq \pi\zeta(x, y, k) + \pi\,\frac{\eta(x, [k+1]y, k)}{ky}$$

Since $v < \pi$ in (5.2)

$$J_3 \leq \pi P\begin{bmatrix} \pi-x, & b' \\ kx & ; b \end{bmatrix}\left\{\overline{\Delta_{x}\left(\frac{\varphi}{uv}\right)}\right\} \qquad \text{[by Th. 6.2 MT (11)]}$$

$$(5.3) \qquad \leq \pi P\begin{bmatrix}\pi-x, & b\\ kx, & ky\end{bmatrix}\left\{\overline{\Delta_{xy}\left(\frac{\varphi}{uv}\right)}\right\} + \pi P\begin{bmatrix}\pi-x, & [k+1]y\\ kx, & ky\end{bmatrix}\left\{\overline{\Delta_{x}\left(\frac{\varphi}{uv}\right)}\right\}$$

applying (9.12) of MT (11) with $a = kx$, $a' = \pi - x$, $b_0 = ky$. We can remove $1/v$ as an internal factor in the last term of (5.3) on noting that $v \geq ky$ therein, adding an external factor of $1/ky$ in accordance with Th. 6.2 MT (11). In the first term in (5.3) one can replace b by $\pi-y \geq b$. Relation (5.2) then follows from (5.3).

The basic lemma will now be proved.

LEMMA 5.2. With $0 < (k+1)x \leq \pi$, $0 < (k+1)y \leq \pi$, $k \geq 1$, $0 < s \leq x$, $0 < t \leq y$, and φ in L over $\begin{bmatrix}\pi, & \pi\\ 0, & 0\end{bmatrix}$

$$(5.4) \qquad \frac{P\begin{bmatrix}kx, & ky\\ 0, & 0\end{bmatrix}(\overline{\varphi})}{\pi(k+1)^4\, xy} - P\begin{bmatrix}\pi, & \pi\\ \pi-x, & \pi-y\end{bmatrix}(\overline{\varphi}).$$

$$\leq \sup_s \zeta(s, y, k) + \sup_s \frac{\eta(s, (k+1)y, k)}{ky} + \sup_t \frac{\xi[kx, t, k]}{kx} .$$

Relation (5.4) is two dimensional. It depends upon two relations essentially one-dimensional in character.[7]

$$(5.5)' \quad P\begin{bmatrix}a', & ky\\ a, & 0\end{bmatrix}[\overline{\varphi}] \leq (k+1)^2 y\left\{\sup_t P\begin{bmatrix}a', & q-t\\ a, & kt\end{bmatrix}\left[\overline{\Delta_t\left(\frac{\varphi}{v}\right)}\right] + \frac{1}{q} P\begin{bmatrix}a', & q\\ a, & q-y\end{bmatrix}[\overline{\varphi}]\right.$$

$$(5.5)'' \quad P\begin{bmatrix}kx, & b'\\ 0, & b\end{bmatrix}[\overline{\varphi}] \leq (k+1)^2 x\left\{\sup_s P\begin{bmatrix}p-s, & b'\\ ks, & b\end{bmatrix}\left[\overline{\Delta_s\left(\frac{\varphi}{u}\right)}\right] + \frac{1}{p} P\begin{bmatrix}p, & b'\\ p-x, & b\end{bmatrix}[\overline{\varphi}]\right\}$$

where $[a,a']$ and $[b,b']$ are arbitrary subintervals of $[0,\pi]$, $0 < (k+1)x \leq p \leq \pi$, $0 < (k+1)y \leq q \leq \pi$, and $0 < s \leq x$, $0 < t \leq y$. It is clear that these two relations differ only in the interchange of the variables x with y, u with v. We shall establish (5.5)'.

PROOF of (5.5)'. One notes the presence of an internal factor $1/v$ in $\overline{\Delta_t\left(\frac{\varphi}{v}\right)}$, with no such factor on the left of (5.5)'. To compensate for the introduction of such a factor in the F-variation of an integral in which v ranges over a closed interval K it is ordinarily sufficient to precede the F-variation by a factor equal to max v over K. [Cf. Th. 6.2, MT (11)]. However, in the case of (5.5)' $\sup 1/v = \infty$ over the interval $K = (0, ky]$.

7. The limits p and q are used in (5.5)" rather than (π, π) [which (p, q) can equal] because of a further use of (5.5) in §11.

We are accordingly led to follow Gergen's procedure of breaking up (0, ky] into the union of a countable number of intervals with end points

$$(5.6) \qquad ky = ky_0 > ky_1 > ky_2 \cdots \rightarrow 0 .$$

Because of the continuity of $\overline{\varphi}$ it is immaterial whether the interval [0,ky] or (0, ky] is used in (5.5)'. Using the latter interval and summing for n = 0, 1, 2, ...

$$(5.7) \quad P\begin{bmatrix} a', ky \\ a\ , 0+ \end{bmatrix} [\overline{\varphi}] \leqq \sum_n P\begin{bmatrix} a', ky_n \\ a\ , ky_{n+1} \end{bmatrix} [\overline{\varphi}] \leqq k \sum y_n \, P\begin{bmatrix} a', ky_n \\ a\ , ky_{n+1} \end{bmatrix} \left[\overline{\tfrac{1}{v}\varphi}\right] = J_4$$

in accordance with Th. 10.2 of MT (11), compensating for the introduction of the internal factor 1/v by the external factor ky_n, [Cf. Th. 6.2, MT (11)]. For the purposes of the proof of (5.5)' the most convenient sequence of points (5.6) is the geometric sequence

$$(5.8) \qquad y_n = y\left(\frac{k}{k+1}\right)^n \qquad [n = 0, 1, \ldots]$$

For later use, we note that $(k+1)y_{n+1} = ky_n$ and that

$$(5.9) \qquad \sum y_n = (k+1)y \qquad [n = 0, 1, \ldots]$$

We shall apply (9.13) of MT (11) to the final variation in (5.7) writing

$$P\begin{bmatrix} a', ky_n \\ a\ , ky_{n+1} \end{bmatrix} \left\{\overline{\tfrac{1}{v}\varphi}\right\} = P\begin{bmatrix} a', b_0+z \\ a\ , b_0 \end{bmatrix} \left\{\overline{\tfrac{1}{v}\varphi}\right\}$$

in (9.13) where $z = k(y_n - y_{n+1}) = y_{n+1}$. If one defines b, b' in (9.13) by setting $b = q - y_{n+1}$, $b' = q$, one verifies that $b' - b = z$. Thus (9.13) can be applied. We find that

$$(5.10) \quad J_4 \leqq k \sum y_n \left\{ P\begin{bmatrix} a', q-y_{n+1} \\ a\ , ky_{n+1} \end{bmatrix} \left\{\overline{\Delta_{y_{n+1}} \left(\tfrac{\varphi}{v}\right)}\right\} + P\begin{bmatrix} a', q \\ a\ , q-y_{n+1} \end{bmatrix} \left\{\overline{\tfrac{1}{v}\varphi}\right\} \right\}$$

$$(5.11) \quad \leqq k(k+1)y \left\{ \sup_t P\begin{bmatrix} a', q-t \\ a\ , kt \end{bmatrix} \left\{\overline{\Delta_t \left(\tfrac{\varphi}{v}\right)}\right\} + \frac{1}{q-y} P\begin{bmatrix} a', q \\ a\ , q-y \end{bmatrix} [\overline{\varphi}] \right.$$

with $0 < t \leqq y$, using (5.9) to sum y_n. In the last term of (5.10) we have first replaced $q - y_{n+1}$ by $q - y$, then removed the internal factor $1/v$, compensating for this by prefixing the factor $\frac{1}{q-y}$. As a consequence of the assumed conditions $k \geqq 1$, $(k+1)y \leqq q$,

$$\frac{k}{q-y} \leqq \frac{k}{q-\frac{q}{k+1}} = \frac{k+1}{q},$$

so that (5.11) implies (5.5)'.

PROOF of 5.4. With $[a,a'] = [0,kx]$ and $q = \pi$, (5.5)' becomes

$$(5.12)\quad P\begin{bmatrix} kx, & ky \\ 0, & 0 \end{bmatrix}[\bar{\varphi}] \leqq (k+1)^2 y \left\{ \sup_t \xi(kx, t, k) + P\begin{bmatrix} kx, & \pi \\ 0, & \pi - y \end{bmatrix}[\bar{\varphi}] \right\} = J_5 .$$

The final variation in (5.12) may be limited as follows. From (5.5)" with $p = \pi$

$$(5.13)\quad P\begin{bmatrix} kx, & \pi \\ 0, & \pi - y \end{bmatrix}[\bar{\varphi}] \leqq (k+1)^2 x \left\{ \sup_s P\begin{bmatrix} \pi - s, & \pi \\ ks, & \pi - y \end{bmatrix}\left[\overline{\Delta_s\left(\frac{\varphi}{u}\right)}\right] + P\begin{bmatrix} \pi, & \pi \\ \pi - x, & \pi - y \end{bmatrix}[\bar{\varphi}] \right\}$$

$$(5.14)\quad \leqq \pi (k+1)^2 x \left\{ \sup_s \zeta(s, y, k) + \sup_s \frac{\eta(s, (k+1)y, k)}{ky} + P\begin{bmatrix} \pi, & \pi \\ \pi - x, & \pi - y \end{bmatrix}[\bar{\varphi}] \right\}$$

using (5.2). With the aid of (5.14), $J_5/(k+1)^2 y$ is at most $\sup_t \xi(kx, t, k)$ from (5.12), plus the right member of (5.14).

Relation (5.4) follows.

THEOREM 5.1. $L_p' + L_p'' \Rightarrow C^0$ and $L_2' + L_2'' \Rightarrow C^0$.

This theorem is a corollary of Lemma 5.2. For the terms on the right of (5.4) are $\bar{o}(1)$ when $L_p' + L_p''$ hold, while

$$(5.15)\quad P\begin{bmatrix} \pi, & \pi \\ \pi - x, & \pi - y \end{bmatrix}(\bar{\varphi}) = \bar{o}(1)$$

since φ is in L. Hence the first term in (5.4) is $\bar{o}(1)$. It follows from the definition of $\bar{o}(1)$ that when $\epsilon > 0$ is given, there exists a k_ϵ and then $\delta_\epsilon = \delta(k_\epsilon, \epsilon)$ such that the first member of (5.4) is at most ϵ for $k = k_\epsilon$ and $0 < x < \delta_\epsilon$, $0 < y < \delta_\epsilon$. Thus with $u = k_\epsilon x$, $v = k_\epsilon y$

$$(5.16)\quad P\begin{bmatrix} u, & v \\ 0, & 0 \end{bmatrix}(\bar{\varphi}) \leqq A\,uv \qquad A = \pi \frac{(k_\epsilon + 1)^4}{k_\epsilon^2}\,\epsilon$$

for $0 < u < \delta_0$, $0 < v < \delta_0$ where $\delta_0 = \delta_\epsilon k_\epsilon$. This establishes the theorem.

The conditions L_2 can be written $\zeta(x, y, 1) = o(1)$, $\xi(x, y, 1) = o(x)$, $\eta(x, y, 1) = o(y)$. Upon setting $k = 1$ in (5.4), one sees that $L_2 \Longrightarrow C^0$.

The following theorem is closely related to Th. 5.1.

THEOREM 5.2. (i) $L_P' + L_P'' \Longrightarrow C^A$; (ii) $L_R' + L_R'' + C^0 \Longrightarrow C^A$.

Use will be made of Lemma 3.1 in proving this theorem, setting $f(x) = x$ in Lemma 3.1 and dropping the k from J_k setting

$$(5.17) \qquad J_b^{b'}(x) = P\begin{bmatrix} x, & b' \\ 0, & b \end{bmatrix} (\overline{\varphi}) \qquad (b'-b = y)$$

where $[b,b']$ is an arbitrary subinterval of $[0,\pi]$. Observe that $J_0^y(x) = O(x\, y)$, by virtue of Th. 5.1 in case (i), and by hypothesis in case (ii). Thus condition (3.7) of Lemma 3.1 is satisfied in both cases. Note that

$$P\begin{bmatrix} x, & b' \\ 0, & b \end{bmatrix} [\overline{\varphi}] \leqq \pi P\begin{bmatrix} x, & b' \\ 0, & b \end{bmatrix} \left[\overline{\tfrac{1}{v}\varphi}\right] \qquad \text{[Th. 6.2. MT (11)]}$$

To show that condition (3.6) of Lemma 3.1 is satisfied in case (i) we use (9.12), MT (11), setting $a = 0$, $a' = x$, $b_0 = ky$, $\omega = \varphi/v$ therein. Whence for $0 < ky < b$

$$P\begin{bmatrix} x, & b' \\ 0, & b \end{bmatrix} \left(\overline{\tfrac{1}{v}\varphi}\right) \leqq P\begin{bmatrix} x, & b \\ 0, & ky \end{bmatrix} \left[\overline{\Delta_y\left(\tfrac{\varphi}{v}\right)}\right] + P\begin{bmatrix} x, & (k+1)y \\ 0, & ky \end{bmatrix} \left[\overline{\tfrac{1}{v}\varphi}\right]$$

$$\leqq \xi(x, y, k) + \frac{1}{ky} P\begin{bmatrix} x, & (k+1)y \\ 0, & 0 \end{bmatrix} (\overline{\varphi}) \leqq \bar{o}(x)$$

since (5.1)" and C^0 hold. [Cf. Th. 5.1]. Thus condition (3.6) of Lemma 3.1 is satisfied in case (i). To establish condition (3.6) in case (ii) we again use (9.12), MT (11), setting $a = 0$, $a' = x$, $b_0 = ky$ and $\omega = \varphi$. Then for $0 < ky < b$

$$P\begin{bmatrix} x, & b' \\ 0, & b \end{bmatrix} (\overline{\varphi}) \leqq P\begin{bmatrix} x, & b \\ 0, & ky \end{bmatrix} (\overline{\Delta_y \varphi}) + P\begin{bmatrix} x, & (k+1)y \\ 0, & ky \end{bmatrix} (\overline{\varphi})$$

$$\leqq P\begin{bmatrix} x, & \pi - y \\ 0, & ky \end{bmatrix} (\overline{\Delta_y \varphi}) + P\begin{bmatrix} x, & (k+1)y \\ 0, & 0 \end{bmatrix} (\overline{\varphi})$$

$$\leqq \pi P\begin{bmatrix} x, & \pi - y \\ 0, & ky \end{bmatrix} \left[\overline{\tfrac{1}{v}\Delta_y \varphi}\right] + P\begin{bmatrix} x, & (k+1)y \\ 0, & 0 \end{bmatrix} (\overline{\varphi}) \qquad \text{[Th. 6.2, MT (11)]}$$

$$\leqq \bar{o}(x)$$

from L_R'' and C^0. Thus condition (3.6) is satisfied in case (ii). We conclude that in either case there exist constants x_0, A such that

$$P\begin{bmatrix} x, & y \\ 0, & 0 \end{bmatrix}(\overline{\varphi}) \leq Axy \tag{5.18}$$

for $0 < x < x_0$, $0 < y \leq \pi$. One can similarly establish (5.18) for $0 < y < y_0$, $0 < x \leq \pi$ for suitable y_0 and A. It is then clear that (5.18) holds for suitable A', with $0 < x \leq \pi$, $0 < y \leq \pi$.

Our formulas imply another uniform relation of use later.

LEMMA 5.3. If $L_P' + L_P''$ holds then for suitable x_0, k_0 with $0 < x_0 < \pi$, $0 < k_0$

$$\eta(x, y, k) < Ay \tag{5.19}$$

for $0 < x < x_0$, $k_0 < k$, $0 < y \leq \pi$.

We again make use of Lemma 3.1 setting

$$J_k{}_b^{b'}(x) = P\begin{bmatrix} \pi - x, & b' \\ kx & , b \end{bmatrix}\left\{\overline{\Delta_x\left(\frac{\varphi}{u}\right)}\right\}, \qquad 0 \leq b < b' \leq \pi$$

and $f(x) \equiv 1$ in Lemma 3.1. By L_P'',

$$J_k{}_0^y(x) = \bar{o}(y)$$

so that (3.7) is satisfied. Condition (3.6) is satisfied in the form

$$J_k{}_b^{b'}(x) = \bar{o}(1) \qquad (b'-b = y)\ (0 < b < b' \leq \pi)$$

in accordance with Lemma 5.1. Since

$$J_k{}_0^y(x) = \eta(x, y, k)$$

relation (5.19) follows from (3.8) of Lemma 3.1.

6. <u>The equivalence</u> $L_P = L_R + C^0$. One notes that the characteristic difference between the conditions L_P and L_R is that L_R differences f while L_P differences f/uv, f/u, and f/v. The proof of the principal theorem may be divided as in §5 into a formal part making no references to the order

hypotheses of L_P or L_R and a concluding part in which these hypotheses enter.

<u>The mixed differences</u>. We begin with an identity which expresses $\Delta_{xy}\varphi$ in terms of differences of the L_P type, followed by an identity which expresses $\Delta_{xy}[\varphi/uv]$ in terms of differences of the L_R type. See Gergen (1) p. 51.

$$\frac{\Delta_{xy}(\varphi)}{uv} = \frac{xy}{u^2v^2}(\varphi) + \left(1+\frac{x}{u}\right)\left(\frac{y}{v^2}\right)\Delta_x\left(\frac{\varphi}{u}\right) \tag{6.1}$$

$$+\frac{x}{u^2}\left(1+\frac{y}{v}\right)\Delta_y\left\{\frac{\varphi}{v}\right\} + \left(1+\frac{x}{u}\right)\left(1+\frac{y}{v}\right)\Delta_{xy}\left\{\frac{\varphi}{uv}\right\}$$

$$\Delta_{xy}\left\{\frac{\varphi}{uv}\right\} = \frac{x\,y\,\varphi}{u(u+x)\,v(v+y)} - \frac{y\,\Delta_x\varphi}{(u+x)\,v(v+y)} \tag{6.2}$$

$$-\frac{x\,\Delta_y\varphi}{u(u+x)\,(v+y)} + \frac{\Delta_{xy}\varphi}{(u+x)\,(v+y)}$$

For $k \geqq 1$, $0 < kx \leqq u \leqq \pi - x$, $0 < ky \leqq v \leqq \pi - y$, we have $(1+x/u) \leqq 2$, $(1+y/v) \leqq 2$ so that (6.1) yields the relation [Cf. Th. 6.2, MT (11)]

$$P\left(Q,\left[\overline{\frac{1}{uv}\Delta_{xy}\varphi}\right]\right) \leqq xy\,P\left(Q,\left[\overline{\frac{1}{u^2v^2}\varphi}\right]\right) + 2y\,P\left(Q,\left[\overline{\frac{1}{v^2}\Delta_x\left(\frac{\varphi}{u}\right)}\right]\right) \tag{6.3}$$

$$Q = \begin{bmatrix}\pi - x, & \pi - y\\ kx\;\;, & ky\end{bmatrix}$$

$$+\,2x\,P\left(Q,\left[\overline{\frac{1}{u^2}\Delta_y\left(\frac{\varphi}{v}\right)}\right]\right) + 4\,P\left(Q,\left[\overline{\Delta_{xy}\left(\frac{\varphi}{uv}\right)}\right]\right)$$

It follows from (6.2) that

$$P\left[Q,\left[\overline{\Delta_{xy}\left(\frac{\varphi}{uv}\right)}\right]\right] \leqq xy\,P\left[Q,\left(\overline{\frac{1}{u^2v^2}\varphi}\right)\right] + y\,P\left[Q,\left(\overline{\frac{1}{uv^2}\Delta_x\varphi}\right)\right] \tag{6.4}$$

$$+\;x\,P\left[Q,\left(\overline{\frac{1}{u^2v}\Delta_y\varphi}\right)\right] + P\left[Q,\left(\overline{\frac{1}{uv}\Delta_{xy}\varphi}\right)\right]$$

In deriving this relation we have replaced the factors $1/(u+x)$, $1/(v+y)$ in terms on the right of (6.2) by $1/u$ and $1/v$ respectively in accordance with Th. 6.2 MT (11).

<u>The first differences</u>. We are necessarily concerned with the relations

between the F-variations $\xi(x, y, k)$ and $\eta(x, y, k)$ appearing in L_P'' and the variations

$$(6.5) \qquad \alpha(x, y, k) = P\begin{bmatrix} x, & \pi-y \\ 0, & ky \end{bmatrix}\left\{\overline{\tfrac{1}{v}\Delta_y \varphi}\right\}$$

$$(6.6) \qquad \beta(x, y, k) = P\begin{bmatrix} \pi-x, & y \\ kx, & 0 \end{bmatrix}\left\{\overline{\tfrac{1}{u}\Delta_x \varphi}\right\}$$

appearing in L_R''. The following lemma limits the differences $\eta - \beta$. A similar lemma holds for $\xi - \alpha$.

LEMMA 6.1. For $k \geq 1$, $0 < kx < \pi - x$, $0 < y \leq \pi$

$$(6.7) \quad \frac{|\eta(x,y,k) - \beta(x,y,k)|}{4x} \leq P\begin{bmatrix} \pi, & y \\ (k+1)x, & 0 \end{bmatrix}(\overline{\varphi}) + \int_{(k+1)x}^{\pi} P\begin{bmatrix} u, & y \\ (k+1)x, & 0 \end{bmatrix}(\overline{\varphi})\frac{du}{u^3}.$$

We begin with the formula

$$(6.8) \qquad \Delta_x\left[\frac{\varphi(u,v)}{u}\right] - \frac{1}{u}\Delta_x \varphi(u,v) = -\frac{\varphi(u+x, v)x}{(u+x)u} = \psi_x(u,v)$$

introducing ψ_x. Upon transposing one of the terms on the left of (6.8) to the right and taking the F-variation of the resulting terms over $\begin{bmatrix} \pi-x, & y \\ kx, & 0 \end{bmatrix}$ one infers that

$$(6.9) \quad |\eta(x,y,k) - \beta(x,y,k)| \leq P\begin{bmatrix} \pi-x, & y \\ kx, & 0 \end{bmatrix}(\overline{\psi_x}) = xP\begin{bmatrix} \pi, & y \\ (k+1)x, & 0 \end{bmatrix}\left(\overline{\frac{\varphi}{u(u-x)}}\right) = J_6$$

where the last term is obtained from ψ_x on making an x-translation of the interval for u. Since $u \geq (k+1)x \geq 2x$ in the last term in (6.9), we have $u/(u-x) \leq 2$, so that one can replace the internal factor of $1/(u-x)$ in (6.9) by $1/u$ if one compensates with an external factor of 2. See Th. 6.2 MT (11). One then makes use of (6.20)*[8] of MT (11) setting $f(u) = 1/u^2$ therein. Since $f(\pi) = 1/\pi^2$ this gives

$$J_6 \leq \frac{2x}{\pi^2} P\begin{bmatrix} \pi, & y \\ (k+1)x, & 0 \end{bmatrix}(\overline{\varphi}) + 4x\int_{(k+1)x}^{\pi} P\begin{bmatrix} u, & y \\ (k+1)x, & 0 \end{bmatrix}(\overline{\varphi})\frac{du}{u^3}$$

and (6.7) follows.

---

8. Equation ( )* designates ( ) with the role of the two arguments (s, t) interchanged.

The next lemma concerns the second terms on the right of (6.3) and (6.4). The relevance of this lemma to the proof that $L_P = L_R + C^0$ is immediate. A similar lemma holds for the third terms on the right of (6.4) and (6.3)

LEMMA 6.2. If $L_R' + L_R''$ holds

$$(6.10) \qquad y\, P\begin{bmatrix}\pi - x, & \pi \\ kx & , ky\end{bmatrix}\left\{\overline{\frac{1}{uv^2}\Delta_x \varphi}\right\} = \bar{o}(1)\ .$$

If $L_P' + L_P''$ holds

$$(6.11) \qquad y\, P\begin{bmatrix}\pi - x, & \pi \\ kx & , ky\end{bmatrix}\left\{\overline{\frac{1}{v^2}\Delta_x \left(\frac{\varphi}{u}\right)}\right\} = \bar{o}(1).$$

The formulas back of this lemma are valid regardless of whether $L_R'+L_R''$ or $L_P'+L_P''$ hold or not. They are as follows. Suppose that $k \geqq 1$, $0 < kx \leqq \pi - x$, $0 < ky \leqq \pi$. From (6.20) of MT (11) with $f(v) = 1/v^2$ and hence $f(\pi) = 1/\pi^2$

$$P\begin{bmatrix}\pi - x, \pi \\ kx & , ky\end{bmatrix}\left\{\overline{\frac{1}{uv^2}\Delta_x \varphi}\right\} \leqq \frac{1}{\pi^2}\, P\begin{bmatrix}\pi - x, \pi \\ kx & , ky\end{bmatrix}\left\{\overline{\frac{1}{u}\Delta_x \varphi}\right\} + 2\int_{ky}^{\pi} P\begin{bmatrix}\pi - x, v \\ kx & , ky\end{bmatrix}\left\{\overline{\frac{1}{u}\Delta_x \varphi}\right\}\frac{dv}{v^3}$$

$$(6.12) \qquad \leqq \frac{\beta(x, \pi, k)}{\pi^2} + 2\int_{ky}^{\pi} \beta(x, v, k)\, \frac{dv}{v^3}\ .$$

With the same generality

$$(6.13) \quad P\begin{bmatrix}\pi - x, \pi \\ kx & , ky\end{bmatrix}\left(\overline{\frac{1}{v^2}\Delta_x\left(\frac{\varphi}{u}\right)}\right) \leqq \frac{\eta(x, \pi, k)}{\pi^2} + 2\int_{ky}^{\pi} \eta(x, v, k)\, \frac{dv}{v^3}\ .$$

If $L_R'+L_R''$ holds it follows from (3.29) that $\beta(x, y, k) \leqq Ay$ for suitable positive k_0, x_0 and $k > k_0$, $0 < x < x_0$, $0 < y \leqq \pi$. Relation (6.10) follows from (6.12). When $L_P' + L_P''$ hold (6.11) similarly follows from (6.13) and Lemma 5.3.

THEOREM 6.1. $L_P \equiv L_R + C^0$.

PROOF that $L_P \Rightarrow L_R + C^0$. One has trivially that $L_P \Rightarrow (C_1)$, and from Th. 5.1, $L_P \Rightarrow C^0$. The basis of the proof that $L_P' + L_P'' \Rightarrow L_R'$ is (6.3) and consists in verifying that each term on the right of (6.3) is $\bar{o}(1)$ provided $L_P' + L_P''$ holds. This is true of the fourth term, which is $4\,\zeta(x,y,k) = \bar{o}(1)$. It is true of the second term by (6.11) and of the third term by (6.11)*. As for the first term in (6.3) recall that $L_P' + L_P'' \Rightarrow C^A$ by Th. 5.2. It follows from Cor. 6.2 of MT (11) that for the interval Q of (6.13) and fixed k

$$P\left[Q, \left(\overline{\frac{\varphi}{u^2v^2}}\right)\right] = \frac{1}{k^2}\, O\left(\frac{1}{xy}\right)$$

so that the first term in (6.3) is $\bar{o}(1)$. Thus $L_P' + L_P'' \Longrightarrow L_R'$;

That $L_P' + L_P'' \Longrightarrow L_R''$ is proved as follows. Since $L_P' + L_P'' \Longrightarrow C^A$ by Th. 5.2 it follows from (6.7) that when $L_P' + L_P''$ holds

$$|\eta(x, y, k) - \beta(x, y, k)| = \bar{o}(y)$$

and hence $\beta(x, y, k) = \bar{o}(y)$. That $\alpha(x, y, k) = \bar{o}(x)$ similarly follows from $(6.7)^*$.

PROOF that $L_R + C^0 \Longrightarrow L_P$. The proof of this is almost the same as that of the inverse implication, using (6.4) in place of (6.3) and using Lemmas 6.1 and 6.2 in the same way. When L_R holds the last term in (6.4) is $\bar{o}(1)$, and the second and third term $\bar{o}(1)$ by (6.10) and $(6.10)^*$. Since $L_R + C^0 \Longrightarrow C^A$ by Th. 5.2, Cor. 6.2 of MT (11) suffices to show that the first term on the right of (6.4) is $\bar{o}(1)$ as above. The proof that $L_R + C^0 \Longrightarrow L_P''$ is by way of (6.7) and $(6.7)^*$, using C^A.

7. Equivalences involving tests of Lebesgue type. The sets of conditions of Lebesgue type are L_1, L_2, L_R, L_P. Some relations have already been obtained. There remains a fundamental lemma.

LEMMA 7.1. (i) $L_R + C^0 \Longrightarrow L_1$; (ii) $L_P + C^0 \Longrightarrow L_2$.

We shall prove (i) in a manner that readily extends to a proof of (ii).

PROOF of (i). The conditions L_1''. Corresponding to the decomposition of $[y, \pi - y]$ into intervals $[y, ky]$ and $[ky, \pi - y]$, with $0 < ky \leq \pi - y$

$$(7.1) \quad P\begin{bmatrix} x, & \pi - y \\ 0, & y \end{bmatrix}\left\{\overline{\frac{1}{v}\Delta_y\varphi}\right\} \leq P\begin{bmatrix} x, & ky \\ 0, & y \end{bmatrix}\left\{\overline{\frac{1}{v}\Delta_y\varphi}\right\} + P\begin{bmatrix} x, & \pi - y \\ 0, & ky \end{bmatrix}\left\{\overline{\frac{1}{v}\Delta_y\varphi}\right\}$$

$$(7.1)' \quad \leq P\begin{bmatrix} x, & (k+1)y \\ 0, & 2y \end{bmatrix}\left(\overline{\frac{1}{v-y}\varphi}\right) + P\begin{bmatrix} x, & ky \\ 0, & y \end{bmatrix}\left(\overline{\frac{1}{v}\varphi}\right) + P\begin{bmatrix} x, & \pi - y \\ 0, & ky \end{bmatrix}\left\{\overline{\frac{1}{v}\Delta_y\varphi}\right\}$$

using the translation principle to get the first term on the right of $(7.1)'$,

$$\leq \frac{1}{y} P\begin{bmatrix} x, & (k+1)y \\ 0, & 2y \end{bmatrix}(\overline{\varphi}) + \frac{1}{y} P\begin{bmatrix} x, & ky \\ 0, & y \end{bmatrix}(\overline{\varphi}) + \alpha(x, y, k)$$

removing the internal factors $1/(v-y)$, and $1/v$. By C^0 and L_R'' this is [Cf. (2.8)]

(7.1)" $\leq x\,\bar{o}(1) + x\,\bar{o}(1) + \bar{o}(x) = \bar{o}(x)$

and this becomes $o(x)$ for the left member of (7.1). The relation (7.1)* runs similarly.

The condition L_1'. For $0 < kx < \pi - x$, $0 < ky < \pi - y$

$$(7.2)\quad P\begin{bmatrix}\pi-x, & \pi-y\\ x & , y\end{bmatrix}\left\{\overline{\frac{1}{uv}\Delta_{xy}\varphi}\right\} \leq P\begin{bmatrix}kx, & \pi-y\\ x, & y\end{bmatrix} + P\begin{bmatrix}\pi-x, & ky\\ x & , y\end{bmatrix} + P\begin{bmatrix}\pi-x, & \pi-y\\ kx & , ky\end{bmatrix}\left\{\overline{\frac{1}{uv}\Delta_{xy}\varphi}\right\}$$

The third term on the right of (7.2) is $\bar{o}(1)$ by L_R'. The first term H_1 on the right of (7.2) is similar to the second. One has

$$H_1 \leq P\begin{bmatrix}(k+1)x, & \pi-y\\ 2x & , y\end{bmatrix}\left\{\overline{\frac{1}{(u-x)v}\Delta_y\varphi}\right\} + P\begin{bmatrix}kx, & \pi-y\\ x, & y\end{bmatrix}\left\{\overline{\frac{1}{uv}\Delta_y\varphi}\right\}$$

$$\leq \frac{1}{x}P\begin{bmatrix}(k+1)x, & \pi-y\\ 0 & , y\end{bmatrix}\left\{\overline{\frac{1}{v}\Delta_y\varphi}\right\} + \frac{1}{x}P\begin{bmatrix}kx, & \pi-y\\ 0, & y\end{bmatrix}\left\{\overline{\frac{1}{v}\Delta_y\varphi}\right\}$$

removing the internal factors $1/(u-x)$ and $1/u$. By virtue of (7.1)" this becomes $\bar{o}(1)$. [Cf. (2.8)].

PROOF of (ii). The proof of (ii) is similar to that of (i), using L_P in place of L_R, $\xi(x, y, k)$ in place of $\alpha(x, y, k)$, and C^0 as in the proof of (i).

The theorem summarizing the relations between the conditions of Lebesgue type follows.

THEOREM 7.1. $L_1 \equiv L_2 \equiv L_P + C^0 \equiv L_R + C^0$; $L_P \equiv L_R + C^0$.

Among relations already established or trivial are the following

$L_1 \Rightarrow L_R + C^0$;	$L_2 \Rightarrow L_P' + L_P''$	[trivial]
$L_2' + L_2'' \Rightarrow C^0$		[by Th. 5.1]
$L_P \equiv L_R + C^0$		[by Th. 6.1]
$L_R + C^0 \Rightarrow L_1$;	$L_P + C^0 \Rightarrow L_2$	[by Lemma 7.1].

Hence

$$(7.3)\qquad L_2 \Rightarrow L_P + C^0 \Rightarrow L_R + C^0 \Rightarrow L_1$$

$$(7.4)\qquad L_1 \Rightarrow L_R + C^0 \Rightarrow L_P + C^0 \Rightarrow L_2$$

Thus $L_1 \equiv L_2$ and one infers the equivalence of the conditions between L_1 and L_2 in (7.4). This establishes Th. 7.1.

8. Proof that $V_Y \to L_2$. Set

$$(8.1) \qquad g(u,v) = \frac{\hat{\Phi}(u,v)}{uv} \qquad \left\{(u,v) \in \begin{bmatrix}\pi, & \pi \\ 0+, & 0+\end{bmatrix} = I\right\}.$$

By definition of V_Y, g satisfies J_H. Recall that a function ω is said to be in FL(I) if ω is in L over every closed subinterval of I and if for some (p,q) in I the integral

$$(8.2) \qquad \bar{\omega}(u,v) = \int_p^u \int_q^v \omega(s,t)\, ds\, dt \qquad \{(u,v) \in I\}$$

is in $\hat{F}(I)$. When ω is in FL(I), $\bar{\omega}$ can be continuously extended over I. [Cf. Lemma 2.1 MT (11)]. So extended $\bar{\omega}$ is in $\hat{F}(\bar{I})$ and is termed an FL-integral. The condition that $\bar{\omega}$ be an FL-integral is independent of the choice of the vertex (p,q) in I. In this section we shall find it convenient to take $(p,q) = (\pi, \pi)$ as a vertex for $\bar{\omega}$. We shall establish the following lemma.

LEMMA 8.1. If φ satisfies V_Y, and g is defined by (8.1), $g_{uv}(u,v)$ exists on a subset E of I such that mE = mI. A function ω such that $\omega(u,v) = g_{uv}(u,v)$ over E, and $\omega(u,v) = 0$ over I - E is in FL(I).

In a brief note Fubini has established the following. There exists subsets E_1 and E_2 of $[0, \pi]$ with $mE_1 = mE_2 = \pi$ and a subset E of I with mI = mE such that the following is true. As a consequence of the fact that φ is in L(I) and that

$$(8.3) \qquad \hat{\varphi}(u,v) = \int_0^u \int_0^v \varphi(s,t)\, ds\, dt$$

$$(8.4)' \qquad \hat{\varphi}_u(u,v) = \int_0^v \varphi(u,t)\, dt \qquad [(u \in E_1)\ (v \in [0, \pi])]$$

$$(8.4)'' \qquad \hat{\varphi}_v(u,v) = \int_0^u \varphi(s,v)\, ds \qquad [(u \in [0, \pi])\ (v \in E_2)]$$

$$(8.4)''' \qquad \hat{\varphi}_{vu}(u,v) = \hat{\varphi}_{uv}(u,v) = \varphi(u,v) \qquad [(u,v) \in E].$$

From these results of Fubini and the definition (8.1) of g we see that $g_{uv}(u,v) = g_{vu}(u,v)$ exists for $(u,v) \in E$. Defining ω as in the Lemma, the explicit form for g_{uv} as derived from (8.1) shows that ω is in L over each closed subinterval of I.

We suppose g extended continuously over I, as is possible. So extended and with v fixed in $(0, \pi]$, $g(\cdot,v)$ with values g(u,v) is absolutely continuous for $u \in [0, \pi]$; for $g(\cdot,v)$ is clearly absolutely continuous for $u \in [e, \pi]$ for each $e > 0$, and in addition is of bounded J-variation for $u \in [0, \pi]$, since g satisfies $\hat{F}$ over $\bar{I}$. A theorem of Lebesgue on absolutely continuous functions of one variable accordingly implies that

$$(8.5)' \qquad g(u,v) = \int_{\pi}^{u} g_u(s,v)ds + g(\pi,v) \qquad \left[(u,v) \in \begin{bmatrix}\pi, & \pi \\ 0, & 0+\end{bmatrix}\right].$$

Similarly

$$(8.5)'' \qquad g(u,v) = \int_{\pi}^{v} g_v(u,t)dt + g(u,\pi) \qquad \left[(u,v) \in \begin{bmatrix}\pi, & \pi \\ 0+, & 0\end{bmatrix}\right].$$

From the formulas (8.4) and (8.1) one infers that $g_u(u,\cdot)$ is absolutely continuous for u fixed in E_1 and $v \in [e, \pi]$, for each $e > 0$. Hence

$$(8.6) \qquad g_u(u,v) = \int_{\pi}^{v} g_{uv}(u,t)dt + g_u(u,\pi), \qquad (u \in E_1),\ (v \in (0, \pi])$$

$$(8.7) \qquad g(u,v) = \int_{\pi}^{u} ds \int_{\pi}^{v} g_{uv}(s,t)dt + \int_{\pi}^{u} g_u(s,\pi)ds + g(\pi,v)$$

for $(u,v) \in I$. On using (8.5)' and (8.2) this gives

$$(8.8) \qquad g(u,v) = \bar{\omega}(u,v) + g(u,\pi) + g(\pi,v) - g(\pi,\pi) \qquad [(u,v) \in I].$$

The form of (8.8) indicates that $\bar{\omega}$ is in $\hat{F}(I)$ with g. This completes the proof of the Lemma.

LEMMA 8.2. For φ, g and ω as in Lemma 8.1 and for a subset $\mathcal{E}$ of I with $m\mathcal{E} = mI$

$$(8.9) \qquad \varphi(u,v) = u\,v\,\omega(u,v) + u\frac{\partial \bar{\omega}}{\partial u}(u,v) + v\frac{\partial \bar{\omega}}{\partial v}(u,v) + g(u,v)$$
$$+ u\,h(u) + v\,k(v) \qquad [(u,v) \in \mathcal{E}]$$

where h and k are integrable over $[0, \pi]$

Set $\mathcal{E} = E \cap [E_1 \times E_2]$. From (8.1), for $(u,v) \in \mathcal{E}$

$$\varphi(u,v) = uv\, g_{uv}(u,v) + u\, g_u(u,v) + v\, g_v(u,v) + g(u,v). \tag{8.10}$$

Making use of (8.6) and (8.6)*, (8.10) implies that (8.9) holds for $(u,v) \in \mathcal{E}$ with

$$h(u) = g_u(u, \pi), \qquad k(v) = g_v(\pi, v),$$

and the lemma follows.

LEMMA 8.3. If h is in $L[0, \pi]$, uh satisfies $L_1 \equiv L_2$.

The condition L_1' is obviously satisfied since $\Delta_y[uh(u)] \equiv 0$. Similarly the condition in L_1'' involving Δ_y is satisfied. We seek then to show that

$$P\begin{bmatrix} \pi - x, & y \\ x & , 0 \end{bmatrix} (\overline{f_x}) = \bar{o}(y), \qquad \text{where } f_x(u) = \frac{1}{u}\Delta_x[u\,h(u)]. \tag{8.11}$$

With (π, π) as a vertex of the FL-integral $\bar{f}_x$

$$\bar{f}_x(u) = \int_{\pi}^{v} dt \int_{\pi}^{u} f_x(s)\, ds = (v - \pi) \int_{\pi}^{u} f_x(s)\, ds .$$

Thus $\bar{f}_x(u)$ is the product of a function of v by a function of u, and has an F-variation which is the product of the J-variations of its factors. Thus

$$P\begin{bmatrix} \pi - x, & y \\ x & , 0 \end{bmatrix} \bar{f}_x(u) = T_0^y (v - \pi)\, T_x^{\pi - x} \int_{\pi}^{u} f_x(s)\, ds = y \int_x^{\pi - x} |f_x(s)|\, ds .$$

To establish (8.11) it is sufficient to show that

$$\int_x^{\pi - x} |f_x(s)|\, ds = \bar{o}(1). \tag{8.12}$$

The condition (8.12) is precisely the 1-dimensional condition of Lebesgue on uh, analogous to L_1 and is satisfied since uh satisfies the 1-dimensional Dini condition. [Cf. Hardy (2)]. Finally the condition C^0 in L_1 is satisfied in the form

$$P\begin{bmatrix} x, & y \\ 0, & 0 \end{bmatrix} (\widehat{uh}) = y \int_0^x |sh(s)|\, ds = o(xy).$$

This establishes the lemma.

Conditions ${}_0L_2$. Under the conditions L_2 on φ we have supposed that φ is in L over $I = \begin{bmatrix} \pi, & \pi \\ 0+, & 0+ \end{bmatrix}$. If φ is in L merely over each closed subinterval of I, L_2' is still well defined and will be denoted by ${}_0L_2'$, while conditions ${}_0L_2''$ of the form

$$P\begin{bmatrix} x, & \pi-y \\ 0+, & y \end{bmatrix}\left\{\overline{\Delta_y\left(\frac{\varphi}{v}\right)}\right\} = o(x) \qquad P\begin{bmatrix} \pi-x, & y \\ x & , 0+ \end{bmatrix}\left\{\overline{\Delta_x\left(\frac{\varphi}{u}\right)}\right\} = o(y)$$

are well defined. We shall find the conditions ${}_0L_2 = {}_0L_2' + {}_0L_2''$ useful in the proof of Th. 8.1. When conditions ${}_0L_2''$ are <u>satisfied</u> the 0+ in these conditions can be replaced by 0 in accordance with Lemma 2.1 of MT (11).

THEOREM 8.1. $V_Y \Rightarrow L_2$.

To prove this theorem use will be made of the representation (8.9) of φ. We shall show that the functions defined by the respective terms on the right of (8.9) satisfy ${}_0L_2$. This is true of uh by Lemma 8.3, and similarly true of vk. It is true of g since g satisfies J_H by definition of V_Y and since $J_H \Rightarrow L_1$ as we shall show in §9.

PROOF <u>that</u> $uv\omega$ <u>satisfies</u> ${}_0L_2$. The condition ${}_0L_2'$ is satisfied by $uv\omega$ since

$$P\begin{bmatrix} \pi-x, & \pi-y \\ x & , y \end{bmatrix} \quad (\overline{\Delta_{xy}\omega}) = o(1) \qquad \text{[by (8.5)'' MT (11)].}$$

The first condition in ${}_0L_2''$ is satisfied by $uv\omega$, since $u\omega$ is in FL(I) with ω and

$$P\begin{bmatrix} x, & \pi-y \\ 0+, & y \end{bmatrix}\left\{\overline{\Delta_y(u\,\omega)}\right\} \quad \leq \quad x\,P\begin{bmatrix} x & , \pi-y \\ 0+, & y \end{bmatrix}(\overline{\Delta_y\omega}) = o(x)$$

on removing the internal factor u, compensating with the external factor x [Th. 6.2, MT (11)] and then using (8.5)' MT (11). The remaining condition in ${}_0L_2''$ is similarly satisfied by $uv\omega$.

PROOF <u>that</u> $u\frac{\partial\bar\omega}{\partial u}$ <u>satisfies</u> ${}_0L_2'$. This condition takes the form

$$J = P\begin{bmatrix} \pi-x, & \pi-y \\ x & , y \end{bmatrix}\left\{\overline{\Delta_{xy}\psi}\right\} = o(1) \qquad \left[\text{where } \psi = \frac{1}{v}\frac{\partial\bar\omega}{\partial u}\right]. \tag{8.13}$$

To establish (8.13) let b' be a constant with $0 < b' < \pi$ and take y so that $0 < 3y < b' < \pi$. Split the interval $[y, \pi-y]$ into the intervals $[y, b'-y]$ and $[b'-y, \pi-y]$ and note that

$$(8.14) \qquad J \leq J_1 + J_2 = P\begin{bmatrix}\pi-x, & \pi-y \\ x, & b'-y\end{bmatrix}\left\{\overline{\Delta_{xy}\psi}\right\} + P\begin{bmatrix}\pi-x, & b'-y \\ x, & y\end{bmatrix}\left\{\overline{\Delta_{xy}\psi}\right\}$$

Let J_2 be decomposed still further setting $\psi(u+x,v) = \theta_x(u,v)$. Then

$$(8.15) \qquad J_2 \leq P\begin{bmatrix}\pi-x, & b'-y \\ x, & y\end{bmatrix}\left\{\overline{\Delta_y \theta_x}\right\} + P\begin{bmatrix}\pi-x, & b'-y \\ x, & y\end{bmatrix}\left\{\overline{\Delta_y \psi}\right\}$$

$$\leq 2 P\begin{bmatrix}\pi, & b'-y \\ x; & y\end{bmatrix}\left\{\overline{\Delta_y \psi}\right\} \qquad \text{[using translation principle]}$$

$$(8.16) \qquad \leq 4 \log 2 \, P\begin{bmatrix}\pi, & b' \\ x, & 0\end{bmatrix} (\overline{\omega}) \qquad \text{[by (8.21) MT (11)]}$$

Let $e > 0$ be prescribed. It follows from (8.16) and Cor. 3.5 MT (11) that $J_2 < e$ if $b' > 0$ be sufficiently small, say $b' = 3\beta$. The derivation of (8.21) MT (11) upon which (8.16) depends, was subject to the condition $0 < 3y < b'$. We now admit y only if $0 < y < \beta$.

We turn to J_1 in (8.14). We shall see that ψ is in FL(Q) for $Q = (0,\pi] \times [\beta,\pi]$. In fact, with (π,π) the vertex of $\overline{\psi}$,

$$\overline{\psi}(u,v) = \int_\pi^u \int_\pi^v \frac{1}{t} \frac{\partial \overline{\omega}}{\partial u}(s,t)\, ds dt = \int_\pi^v \frac{1}{t} \overline{\omega}(u,t) dt \qquad \text{[Cf. (8.2)]}$$

and it follows from (5.23) of MT (11) that

$$P(Q,\overline{\psi}) \leq \frac{\pi}{\beta} P(I,\overline{\omega}) < \infty \qquad \left\{ I = \begin{bmatrix}\pi, & \pi \\ 0+; & 0+\end{bmatrix} \right\}.$$

Thus ψ is in FL(Q). For $b' = 3\beta$ and $0 < y < \beta$, one has $b'-y > \beta$, so that

$$J_1 \leq P\begin{bmatrix}\pi-x, & \pi-y \\ 0, & \beta\end{bmatrix}\left\{\overline{\Delta_{xy}\psi}\right\}.$$

It follows from (8.5)″ of MT (11) that $J_1 < e$ if $x > 0$ and $y > 0$ are sufficiently small. For such x, and y, $J_1 + J_2 < 2e$. Thus (8.13) holds.

PROOF <u>that</u> $u \dfrac{\partial \overline{\omega}}{\partial u}$ <u>satisfies</u> ${}_0L_2''$. The first condition in ${}_0L_2''$ is satisfied since

$$P\begin{bmatrix}x, & \pi-y \\ 0+, & y\end{bmatrix} \left\{\overline{\Delta_y\left(\frac{u}{v}\frac{\partial \overline{\omega}}{\partial v}\right)}\right\} \leq x\, P\begin{bmatrix}x, & \pi-y \\ 0+, & y\end{bmatrix}\left\{\overline{\Delta_y\left(\frac{1}{v}\frac{\partial \overline{\omega}}{\partial v}\right)}\right\} \qquad \text{[by Th. 6.2 MT(11)]}$$

$$\leq 2x \log 2 \, P\begin{pmatrix}x, & \pi \\ 0; & 0\end{pmatrix} (\overline{\omega}) \qquad \text{[by Th. 8.3 MT(11)]}$$

$$\leq 2x\ o(1) \qquad \text{[by Cor. 3.1 MT(11)]}$$

The second condition in ${}_0L_2''$ is satisfied since

$$P\begin{bmatrix}\pi-x, & y\\ x & , 0\end{bmatrix}\left\{\overline{\Delta_x \frac{\partial\bar\omega}{\partial u}}\right\} \leq y\ P\begin{bmatrix}\pi-x, & y\\ x & , 0\end{bmatrix}\left(\overline{\Delta_x\omega}\right) = y\ o(1)$$

by (8.6) MT (11) and $(8.5)'^{*}$ MT (11).

Thus $u\frac{\partial\bar\omega}{\partial u}$ satisfies ${}_0L_2$. The proof that $v\frac{\partial\bar\omega}{\partial v}$ satisfies ${}_0L_2$ is similar. Each term in the sum (8.9) representing φ thus satisfies ${}_0L_2$ if φ satisfies V_Y and the proof of the theorem is complete since φ then satisfies L_2.

9. Concluding implications. If A and B are MT-conditions we say that an implication $A \Rightarrow B$ is reduced relative to MT-conditions if the relations

$$A \Rightarrow C \Rightarrow B \tag{9.1}$$

for an MT-condition C implies that $A \equiv C$ or else $C \equiv B$. The following theorem summarizes the implications other than those between conditions of Lebesgue type.

THEOREM 9.1. The following implications

(a) $J_H \Rightarrow Y \Rightarrow Y_P \Rightarrow L_P$

(b) $J_H \Rightarrow Y \Rightarrow L_1$

(c) $J_H \Rightarrow V_Y \Rightarrow L_1$

(d) $D_Y \Rightarrow V_Y \Rightarrow L_1$

are reduced relative to MT-conditions provided[9] $Y_P \not\Rightarrow L_1$. If $Y_P \Rightarrow L_1$, (a) and (b) are to be replaced by the implications

(e) $J_H \Rightarrow Y \Rightarrow Y_P \Rightarrow L_1 \Rightarrow L_P$

which are reduced relative to MT-conditions.

9. We shall see in §10 that $L_1 \not\equiv Y_P$. It is possible that $Y_P \Rightarrow L_1$ as far as we know. For the corresponding 1-dimensional tests we have shown that $Y_P \Rightarrow L_1$. [Cf. MT (10)].

(a) PROOF that $J_H \to Y$. We assume that φ satisfies J_H and seek to prove that φ satisfies $Y = Y' + (C_0)$. By definition, Y' is the condition

$$P\begin{bmatrix} x, & y \\ 0, & 0 \end{bmatrix}(uv\,\varphi) \leq Axy \qquad \left\{(x,y) \in \begin{bmatrix} \pi, & \pi \\ 0+, & 0+ \end{bmatrix} = I\right\} \tag{9.2}$$

Under J_H, φ is in $\hat{F}$ over I. Without loss of generality we can suppose φ modified on the coordinate axes so that

$$\varphi(0+,0+) = \varphi(0,0); \qquad \varphi(u,0+) = \varphi(u,0); \qquad \varphi(0+,v) = \varphi(0,v)$$

for $(0 \leq u \leq \pi)$ and $(0 \leq v \leq \pi)$. These limits exist since φ is in $\hat{F}(I)$. With φ so modified the quadriant limit $\varphi(0+,0+) = \varphi(0,0+) = \varphi(0+,0)$. The modification of φ (if any) affects the satisfaction neither of J_H nor of Y. When φ has these boundary values $P(\bar{I}, \varphi) = F(I, \varphi)$. [See Lemma 5.1 MT (5)]. Under the conditon (C_0) contained in J_H. $\varphi(0+,0+) = 0$.

To continue set

$$\varphi_1(u) = \varphi(u,0) \qquad \varphi_2(v) = \varphi(0,v)$$

$$\varphi_3(u,v) = \varphi(u,v) - \varphi_1(u) - \varphi_2(v)$$

Observe that

$$\varphi_3(0,v) = \varphi_3(u,0) = 0. \qquad (0 \leq u \leq \pi)(0 \leq v \leq \pi). \tag{9.3}$$

Since $P(\bar{I}, \varphi) = P(\bar{I}, \varphi_3) < \infty$ and (9.3) holds it follows from Th. 4.1 of MT (11), on setting $h(u) = u$ and $k(v) = v$ therein, that

$$P\begin{bmatrix} x, & y \\ 0, & 0 \end{bmatrix}(uv\,\varphi_3) \leq A_1\, xy \qquad [(x,y) \in I] \tag{9.4}$$

for a suitable constant A_1. Since $T_0^{\pi}(\varphi_2)$ is finite and $\varphi_2(0) = 0$ it follows from Lemma 4.1, MT (11) on setting $k(v) = v$ therein, that

$$T_0^{y}(v\,\varphi_2) \leq A_2\, y \qquad (0 < x \leq \pi)$$

for a suitable constant A_2. Hence

$$P\begin{bmatrix} x, & y \\ 0, & 0 \end{bmatrix}(uv\,\varphi_2) \leq A_2\, x\, y \tag{9.5}$$

Similarly

$$P\begin{bmatrix}x, & y\\ 0, & 0\end{bmatrix}(uv\varphi_1) \leq A_3xy \qquad [(x,y)\in I] \tag{9.6}$$

Since $\varphi(u,v) = \varphi_1(u) + \varphi_2(u) + \varphi_3(u,v)$ (9.4), (9.5) and (9.6) imply (9.2).

This completes the proof that $J_H \Rightarrow Y$.

(a) PROOF that $Y \Rightarrow Y_P \Rightarrow L_P$. That $Y \Rightarrow Y_P$ is trivial, while $Y_P \Rightarrow L_P$ by Cor. 4.1.

(b) PROOF that $Y \Rightarrow L_1$. We have

$$Y = Y' + (C_0) \Rightarrow L_P' + L_P'' \text{ (Th. 4.1)} + C^0\text{(trivial)} \Rightarrow L_1 \text{ (Th. 7.1)}$$

(c) PROOF that $J_H \Rightarrow V_Y$. According to (7.27) of MT (11)

$$P\left[I, \frac{1}{uv}\hat{\varphi}\right] \leq P[I, \varphi] \qquad \left\{I = \begin{bmatrix}\pi, & \pi\\ 0+, & 0+\end{bmatrix}\right\} \tag{9.7}$$

Since $P[I, \varphi] < \infty$ under J_H, the left member of (9.7) is finite. Since φ satisfies (C_0) under J_H, $\varphi(u,v) = o(1)$ and $\hat{\varphi}(u,v) = o(uv)$. It follows that $\hat{\varphi}(u,v)/uv = o(1)$ and so satisfies (C_0). Thus $\hat{\varphi}/uv$ satisfies J_H so that $J_H \Rightarrow V_Y$.

PROOF that $D_Y \Rightarrow V_Y$. Set $g(u,v) = \hat{\varphi}(u,v)/uv$ over $I = \begin{bmatrix}\pi, & \pi\\ 0+, & 0+\end{bmatrix}$. Assuming that φ satisfies D_Y we seek to prove that φ satisfies V_Y, that is that g satisfies J_H. By virtue of Th. 7.1 MT (11)

$$P\left(I, \frac{1}{uv}\hat{\varphi}\right) \leq 4\,P\left(I, \left(\overline{\frac{1}{uv}\varphi}\right)\right) < \infty \tag{9.8}$$

since the right member of (9.8) is finite under D_Y. We continue by proving the following.

(a) <u>For each</u> $b \in (0, \pi]$ $g(0+,b) = 0$.

By definition of D_Y, $\overline{\varphi/uv}$ is in $\hat{F}(I)$. The relations

$$P\left(I, \left\{\overline{\tfrac{1}{u}\varphi}\right\}\right) = P\left(I, \left\{\overline{v\left(\tfrac{\varphi}{uv}\right)}\right\}\right) \leq \pi P\left(I, \left\{\overline{\tfrac{1}{uv}\varphi}\right\}\right) < \infty$$

imply that $(\overline{\frac{1}{u}\varphi})$ is in $\hat{F}(I)$. Set $Q = \begin{bmatrix}\pi, & \pi\\ 0+, & 0\end{bmatrix}$. <u>Since</u> φ/u is <u>in L</u> over every closed subinterval of Q, and since $P\left(Q, |\overline{\tfrac{1}{u}\varphi}|\right) = P\left(I, |\overline{\tfrac{1}{u}\varphi}|\right) < \infty$ [Cf. Lemma 5.1 MT (5)], φ/u is in <u>FL(Q)</u> by definition of FL(Q). Using the "vertex" $(\pi, 0)$ in Q set $\gamma(u,v) = (\overline{\frac{1}{u}\varphi})$. The FL-integral γ is defined over $\bar{Q}$ by continuous extension of its values on Q. For $(u,v) \in Q$

$$(9.9) \qquad \gamma(u,v) = \int_{\pi}^{u} \frac{ds}{s} \int_{0}^{v} \varphi(s,t)\, dt.$$

For $0 < b \leq \pi$ $\gamma(\cdot,b)$ is absolutely continuous over $[e,\pi]$ for each $e > 0$, and of bounded J-variation over $[0,\pi]$ since γ is in $\hat{F}(\bar{I})$. Hence $\gamma(\cdot,b)$ is absolutely continuous over $[0,\pi]$. From the definition of g and γ and with $0 < u \leq \pi$

$$(9.10) \qquad g(u,b) = \frac{1}{u} \int_{0}^{u} f(s)\, ds \qquad [\text{where } f(s) = \frac{1}{b} \int_{0}^{b} \varphi(s,t)\, dt]$$

$$(9.11) \qquad \gamma(u,b) = \int_{\pi}^{u} k(s)\, ds \qquad [\text{where } k(s) = \int_{0}^{b} \frac{\varphi(s,t)}{s}\, dt].$$

Note that $k(s)/b = f(s)/s$ for $s \in (0,\pi]$. Since $\gamma(\cdot,b)$ is absolutely continuous over $[0,\pi]$, k and hence f/s is integrable over $[0,\pi]$. Hence

$$\lim_{u \to 0} \int_{0}^{u} \frac{|f(s)|}{s}\, ds = 0 .$$

Then since

$$|g(u,b)| = \left|\frac{1}{u} \int_{0}^{u} f(s) ds\right| \leq \int_{0}^{u} \frac{|f(s)|}{s}\, ds$$

we conclude that $g(0+,b) = 0$ for $b \in (0,\pi]$. This establishes (a).

Similarly $g(a,0+) = 0$ for $a \in (0,\pi]$. We have seen in (9.8) that $P(I,g) < \infty$. It follows from the proof of Lemma 5.1 MT (5) that g admits an extension g^e over $(\bar{I})$ which vanishes when $u = 0$ or $v = 0$. Since g is continuous over I the existence of the quadrant limit of g from the right of the points of $\bar{I}$ at which $u = 0$ or $v = 0$ implies that g^e is continuous at these points. Hence $g^e(0,0) = 0$. The vanishing of g^e on the axes together with the relation

$$P(\bar{I},g^e) = P(I,g)$$

of Lemma 5.1 MT (5) implies that g^e is in $\hat{F}(\bar{I})$. Hence g is in $\hat{F}(I)$ and $g(0+,0+) = 0$. Thus g satisfies J_H.

This completes the proof of the theorem except for the statement that the implications are reduced; this will follow with the aid of the results of section 10.

<u>10</u>. <u>The denial of implications</u>. <u>Examples of product type</u>. We compare 2-dimensional tests with 2-dimensional tests and never with 1-dimensional

tests. If A represents a test let K^A denote the class of functions which satisfy A. If A and B are two tests and $K^A \supset K^B$ and $K^A \not\subset K^B$ we say that B is <u>below</u> A and A <u>above</u> B. Thus L_1 is above J_H. If A is neither above nor below B we say that A and B are <u>incomparable</u>. Thus J_H and D_Y are incomparable as we shall see.

As in the introduction we shall refer to the 2-dimensional tests listed by Gergen as of <u>classical type</u> and the corresponding new tests introduced in this paper as of MT-type. We shall be concerned with comparisons between corresponding tests of classical and MT-types and comparisons between two tests of MT-types.

The denial of implications, for example, $L_1 \not\Rightarrow J_H$, $L_1 \not\Rightarrow (L_1)$, may be inferred from examples of functions φ of two types: those of <u>product type</u> [i.e. $\varphi(u,v) = h(u)\, k(v)$], and those not of product type. The importance of examples of product type arises from the fact that when h is defined over an interval U and k over an interval V, and hk has the values $h(u)\, k(v)$ then

$$V(U \times V,\ hk) = P(U \times V,\ hk) = T(U,h)\, T(V,k) \tag{10.1}$$

Those implications between two tests of classical type or MT-type which we are able to deny by means of product functions either involve two tests below those of Lebesgue type, for example $J_H \not\Rightarrow D_Y$, or involve a test of Lebesgue type and one not of Lebesgue type, for example $L_1 \not\Rightarrow Y_P$. The only denial of implications between tests of Lebesgue type, for example $L_R \not\Rightarrow (L_R)$ are (as far as we know) by way of functions not of product type as described in §§11,12.

In order to take full advantage of functions of product type we must clarify the terminology of 1-dimensional tests. A bar superimposed above the generic symbol for a test will indicate that it is 1-dimensional. Thus (J_H) is the classical Hardy-Jordan 2-dimensional test, J_H the new test of of MT-type, and $\bar{J}$ the 1-dimensional test.

The 1-dimensional tests are concerned with an even-function $f \in L[0,\pi]$, a constant s, and the function ψ, with values $\psi(u) = f(u) - s$. Set

$$\hat{\psi}(u) = \int_0^u \psi(t)\, dt \qquad (0 \leq u \leq \pi)$$

The condition $\psi(u) = o(1)$ and $\hat{\psi}(x) = o(x)$ will be denoted by $\bar{C}_0$ and $\bar{C}_1$ respectively. The condition

$$\int_0^x |\psi(t)|\, dt = o(x) \qquad (x > 0)$$

will be denoted by $\bar{C}^0$. With this understood the 1-dimensional tests to which we shall refer are defined below, listing the conditions on ψ opposite the symbol for the test

$\bar{J}$: $T_{0+}^{\pi}(\psi) < \infty$ and $\bar{C}_0$

$\bar{V}$: $\hat{\psi}/u$ satisfies $\bar{J}$

$\bar{D}$: $\psi/u \in L[0, \pi]$

$\bar{Y}$: $T_0^x(u\psi) \leqq Ax$ and $\bar{C}_0$

$\bar{Y}_P$: $T_0^x(u\psi) \leqq Ax$ and $\bar{C}_1$

$\bar{L}_1$: $\int_x^{\pi} \left| \frac{\psi(t+x) - \psi(t)}{t} \right| dt = o(1)$ and $\bar{C}^0$.

THEOREM 10.1. Figure 3 gives a valid representation of relations of inclusion and incomparability between 1-dimensional tests.

The correctness of Figure 3 is a consequence of the following implications

(10.2) $\bar{J} \Longrightarrow \bar{Y} \Longrightarrow \bar{L}_1$; $\quad \bar{J} \Longrightarrow \bar{V} \Longrightarrow \bar{L}_1$; $\quad \bar{D} \Longrightarrow \bar{V}$

and of the following denial of implications,

(10.3) $\bar{J} \not\Longrightarrow \bar{D}$; $\quad \dot{\bar{Y}} \not\Longrightarrow \bar{V}$; $\quad \bar{D} \not\Longrightarrow \bar{Y}_P$

proofs of all of which may be inferred from Hardy (2), taken with the trivial implication $\bar{Y} \Longrightarrow \bar{Y}_P$, the relation $\bar{Y}_P \not\Longrightarrow \bar{Y}$ shown by an example in the last paragraph of Hardy and Littlewood, and finally the new theorem, established in MT (10)

THEOREM 10.2 $\bar{Y}_P \Longrightarrow \bar{L}_1$.

A necessary and sufficient condition that two tests represented in Figure 3 should be incomparable is that their semi-circles should intersect. The relations of incomparability indicated in Figure 3 are easily derived from the above relations. For example, $\bar{J}$ and $\bar{D}$ are incomparable. Otherwise $\bar{D} \Longrightarrow \bar{J} \Longrightarrow \bar{Y} \Longrightarrow \bar{Y}_P$ contrary to (10.3). Similarly $\bar{Y}$ and $\bar{D}$ are incomparable as well as $\bar{V}$ and $\bar{Y}$.

We shall prove the following theorem.

THEOREM 10.3. Figure 1 gives a valid representation of relations between MT-tests.

As proof recall that the implications (10.2) have their counterparts

(10.4) $\qquad J_H \Rightarrow Y \Rightarrow L_1; \qquad J_H \Rightarrow V_Y \Rightarrow L_1; \qquad D_Y \Rightarrow V_Y$

established in the earlier sections. Moreover we shall see that

(10.5) $\qquad J_H \not\Rightarrow D_Y; \qquad Y \not\Rightarrow V_Y; \qquad D_Y \not\Rightarrow Y_P$.

PROOF that $J_H \not\Rightarrow D_Y$. Let ψ with values $\psi(u)$ satisfy $\bar{J}$ but not $\bar{D}$, as is possible since $\bar{J} \not\Rightarrow \bar{D}$. Let φ then have the values $\varphi(u,v) = \psi(u)\,\psi(v)$. One sees that φ satisfies J_H but not D_Y.

PROOF that $Y \not\Rightarrow V_Y$. Let ψ with values $\psi(u)$ satisfy $\bar{Y}$ but not $\bar{V}$, as is possible since $\bar{Y} \not\Rightarrow \bar{V}$. Let $\varphi(u,v) = \psi(u)\ \psi(v)$. One sees that φ satisfies Y. Observe that

(10.6) $$\frac{\hat{\varphi}(u,v)}{uv} = \frac{\hat{\psi}(u)}{u}\ \frac{\hat{\psi}(v)}{v} \qquad (u > 0,\ v > 0).$$

By hypothesis $\hat{\psi}/u$ fails (Case I) to satisfy $\bar{C}_0$, or else (Case II) $T^{\pi}_{0+}(\hat{\psi}/u) = \infty$. In Case I

$$\frac{\hat{\varphi}(u,u)}{u^2} = \left[\frac{\hat{\psi}(u)}{u}\right]^2 \neq o(1) \qquad (u > 0)$$

so that $\hat{\varphi}/uv$ fails to satisfy (C_0). In Case II

$$P\begin{bmatrix}\pi, & \pi\\ 0+, & 0+\end{bmatrix}(\hat{\varphi}/uv) = \infty$$

as a consequence of (10.6). In either case φ fails to satisfy V_Y.

PROOF that $D_Y \not\Rightarrow Y$ and $D_Y \not\Rightarrow Y_P$. It is sufficient to show that $D_Y \not\Rightarrow Y_P$ since $Y \Rightarrow Y_P$ trivially. Hardy gives a function ψ which satisfies $\bar{D}$ but for which

$$T^x_0(u\psi) \neq O(x) \qquad (x > 0)$$

If $\varphi(u,v) = \psi(u)\ \psi(v)$

$$P\begin{bmatrix}x, & y\\ 0, & 0\end{bmatrix}(uv\,\varphi) = T^x_0\,(u\psi)\ T^y_0\,(v\psi)$$

so that

$$P\begin{bmatrix}x, & x\\ 0, & 0\end{bmatrix}(uv\,\varphi) = \left[T^x_0\,(u\psi)\right]^2 \neq O(x^2).$$

Hence φ cannot satisfy Y_P.

The correctness of Figure 1 for MT-tests follows from (10.4), (10.5) and the trivial implication $Y \Rightarrow Y_P$.

Figure 2. The implications of the full vectors in Figure 2 are correct by virtue of Th. 9.1 and Th. 7.1. In §1 it is affirmed that no implication between the tests not indicated as doubtful or logically implied by Figure 2 is valid. This is a consequence of Figure 1, of the relation $D_Y \not\Rightarrow Y_P$ just established, and of the relation

$$Y_P \not\Rightarrow Y \tag{10.7}$$

established as follows. In the last paragraph Hardy and Littlewood give a function φ with values $\varphi(u)$ which satisfies $\bar{Y}_P$ but fails to satisfy $\bar{C}_o$, since $\varphi(u) \not\to 0$ as $u \to 0$. One sees that a function with values $\varphi(u)\ \varphi(v)$ will satisfy Y_P but fail to satisfy C_o. Hence (10.7) holds. The validity of Figure 2 can now be checked item by item.

11. Proof that the MT-test $D_Y \not\Rightarrow$ the Gergen test (L_R). From this proof it follows that if X is a test such that $D_Y \to X$, then $X \not\Rightarrow (L_R)$. In particular from our earlier implications and the relation $D_Y \not\Rightarrow (L_R)$ it will follow that

$$D_Y \Rightarrow V_Y \Rightarrow L_1 \Rightarrow L_2 \Rightarrow L_P \Rightarrow L_R, \tag{11.1}$$

and that none of these MT-tests imply (L_R).

To show that $D_Y \not\Rightarrow (L_R)$ we shall need two formal relations. Use will be made of the functions

$$\bar{\alpha}(x, y, k) = V\begin{bmatrix} x, & \pi - y \\ 0, & ky \end{bmatrix}\left(\overline{\tfrac{1}{v}\Delta_y \varphi}\right) \tag{11.2}$$

$$\bar{\beta}(x, y, k) = V\begin{bmatrix} \pi - x, & y \\ kx, & 0 \end{bmatrix}\left(\overline{\tfrac{1}{u}\Delta_x \varphi}\right) \tag{11.3}$$

to establish the following lemma.

LEMMA 11.1. For $0 < (k+1)x < \pi$, $0 < (k+1)y < \pi$ and $0 < s < x$, $0 < t < y$.

$$\frac{V\begin{bmatrix} kx, & ky \\ 0, & 0 \end{bmatrix}(\overline{uv\varphi})}{(k+1)^6\, x^2\, y^2} \qquad \frac{V\begin{bmatrix} (k+1)x, & (k+1)y \\ kx, & ky \end{bmatrix}(\bar{\varphi})}{(k+1)^2\, x\, y}$$

$$\leq \sup_t \frac{\bar{\alpha}[kx, t, k]}{kx} + \sup_s \frac{\bar{\beta}[s, (k+1)y; k]}{(k+1)\, y}. \tag{11.4}$$

The proof of this formal lemma is similar to that of Lemma 5.2. We begin with relations (5.5) altered trivially by the introduction of the factor $1/v$ and $1/u$ in the "integrands" of the last term of (5.5)' and (5.5)" respectively, and further altered by replacing P by V. Thus

$$(11.5)' \quad V\begin{bmatrix} a', ky \\ a, 0 \end{bmatrix}(\overline{\varphi}) \leq (k+1)^2 y \left\{ \sup_t V\begin{bmatrix} a', q-t \\ a, kt \end{bmatrix} \left\{\overline{\Delta_t\left(\frac{\varphi}{v}\right)}\right\} + V\begin{bmatrix} a', q \\ a, q-y \end{bmatrix}\left(\overline{\frac{1}{v}\varphi}\right)\right.$$

$$(11.5)'' \quad V\begin{bmatrix} kx, b' \\ 0, b \end{bmatrix}(\overline{\varphi}) \leq (k+1)^2 x \left\{ \sup_s V\begin{bmatrix} p-s, b' \\ ks, b \end{bmatrix} \left\{\overline{\Delta_s\left(\frac{\varphi}{u}\right)}\right\} + V\begin{bmatrix} p, b' \\ p-x, b \end{bmatrix}\left(\overline{\frac{1}{u}\varphi}\right)\right. .$$

The proof of (11.5) is formally the same as that of (5.5) replacing P by V throughout. Th. 6.2 MT (11) to which reference is made in the proof of (5.5) is immediately obvious when P is replaced by V by virtue of the representation of a Vitali variation $V(Q,\bar{g})$ in the form

$$(11.6) \qquad V(Q,\bar{g}) = \int_Q \int |g(s,t)|\, ds\, dt.$$

The reference to Th. 10.2 of MT (11) in the proof of (5.5) is replaced here by the known fact that the integral (11.6) defines a completely additive set function. The reference to (9.13) of MT (11) may here be replaced by a reference to (9.13)a of MT (11). Relations (11.5) follow in this way.

Replace φ by $uv\varphi$ in (11.5)' setting $q = (k+1)y$, $a = 0$, $a' = kx$. In (11.5)" replace φ by $u\varphi$ setting $p = (k+1)x$, $b = ky$, $b' = (k+1)y$. Thus with $0 < t \leq y$, $0 < s \leq x$

$$(11.7)' \quad \frac{V\begin{bmatrix} kx, ky \\ 0, 0 \end{bmatrix} |\overline{uv\varphi}|}{(k+1)^2 y} \leq \sup_t V\begin{bmatrix} kx, (k+1)y-t \\ 0, kt \end{bmatrix} |\overline{\Delta_t(u\varphi)}| + V\begin{bmatrix} kx, (k+1)y \\ 0, ky \end{bmatrix} |\overline{u\varphi}|$$

$$(11.7)'' \quad \frac{V\begin{bmatrix} kx, (k+1)y \\ 0, ky \end{bmatrix} |\overline{u\varphi}|}{(k+1)^2 x} \leq \sup_s V\begin{bmatrix} (k+1)x-s, (k+1)y \\ ks, ky \end{bmatrix} |\overline{\Delta_s\varphi}| + V\begin{bmatrix} (k+1)x, (k+1)y \\ kx, ky \end{bmatrix} |\overline{\varphi}|$$

In the first of these relations replace $\Delta_t(u\varphi)$ by $\frac{1}{v}\Delta_t\varphi$, compensating by a factor $(kx)(k+1)y$ preceding the V. In the second of these relations replace $\Delta_s\varphi$ by $\frac{1}{u}\Delta_s\varphi$ compensating by a factor $(k+1)x$ preceding V. Combining the resulting two relations

$$(11.8)\quad \frac{V\begin{bmatrix} kx, ky \\ 0\ , 0 \end{bmatrix}(\overline{uv\ \varphi})}{(k+1)^2\ y} \leqq (kx)\ (k+1)y\ \sup_t V \begin{bmatrix} kx, (k+1)y-t \\ 0\ ,\ kt \end{bmatrix} \left\{\overline{\tfrac{1}{v}\Delta_t \varphi}\right\}$$

$$+ (k+1)^3\ x^2\ \sup_s V\begin{bmatrix} (k+1)x-s, (k+1)y \\ ks\ \ \ \ \ \ \ \ , ky \end{bmatrix}\left\{\overline{\tfrac{1}{u}\Delta_s \varphi}\right\} + (k+1)^2 x\ V\begin{bmatrix} (k+1)x, (k+1)y \\ kx\ \ \ \ \ , ky \end{bmatrix}(\overline{\varphi}).$$

On recalling that $(k+1)y < \pi$ and $(k+1)x < \pi$ by hypothesis, relation (11.4) follows from (11.8) on dividing (11.8) by $(k+1)^4\ x^2\ y$.

LEMMA 11.2. If φ satisfies (L_R) there exist positive constants a_1, b_1, d, K and $k > 2$ such that for $0 < a < a_1$, $0 < b < b_1$

$$(11.9)\quad \frac{V\begin{bmatrix} a\ , b \\ 0\ , 0 \end{bmatrix}(\overline{uv\ \varphi})}{a^2\ b^2} - \frac{dV\begin{bmatrix} a+\frac{a}{k}\ , b+\frac{b}{k} \\ a\ \ \ \ \ \ , b \end{bmatrix}(\overline{\varphi})}{a\ b} \leqq K.$$

Relation (3.29) holds with P replaced by V, as does its proof on replacing P by V, L_R by (L_R) throughout. The same relation is found in Lemma 6 p. 43 of Gergen (1). From this relation and its conjugate we have

$$(11.10)\quad \overline{\beta}(s,\ (k+1)y,\ k) \leqq A_1\ (k+1)y \qquad [0 < s < s_0,\ k_0 < k,\ 0 < (k+1)y < \pi]$$

$$\overline{\alpha}(kx,\ t,\ k) \leqq A_2\ k\ x \qquad [0 < t < t_0,\ k_0 < k,\ 0 < kx < \pi]$$

for suitably chosen constants s_0, t_0, k_0. Let x and y in (11.4) be subjected to the additional conditions $0 < x < s_0$, $0 < y < t_0$. Then the s and t admitted in (11.4) satisfy the conditions of (11.10) so that both (11.4) and (11.10) hold for $k > k_0$ and imply the relation

$$\frac{V\begin{bmatrix} kx\ , ky \\ 0\ \ , 0 \end{bmatrix}|\overline{uv\ \varphi}|}{(k+1)^6\ x^2y^2} - \frac{V\begin{bmatrix} (k+1)x\ , (k+1)y \\ kx\ \ \ \ \ \ \ ,\ \ ky \end{bmatrix}(\overline{\varphi})}{(k+1)^2\ x\ y} \leqq A_1 + A_2\ .$$

If one fixes $k > k_0$ and sets $a = kx$, $b = ky$, (11.9) holds for $0 < a < k\ s_0$, $0 < b < k\ t_0$ and for obvious choices of d and K, supposing $(k+1)s_0 < \pi$, $(k+1)t_0 < \pi$.

THEOREM 11.1. $D_Y \not\Rightarrow (L_R)$.

It follows from Lemma 7.1 of MT (7) that there exists a function φ

in L over $\begin{bmatrix}\pi, \pi \\ 0+, 0+\end{bmatrix} = I$ and such that $P\left[I, \left(\overline{\frac{1}{uv}\varphi}\right)\right] < \infty$, and for suitably chosen arbitrarily small constants $c > 0$

$$(11.11) \qquad V\begin{bmatrix}c, c \\ 0, 0\end{bmatrix} |\overline{uv\varphi}| \geq c^3 \qquad\qquad V\begin{bmatrix}2c, 2c \\ c, c\end{bmatrix} (\overline{\varphi}) = 0 .$$

Such a φ satisfies D_Y by definition of D_Y. Such a φ cannot satisfy (L_R). For if it did (11.9) would hold for this φ and for each $a = b = c < \min [a_1, b_1]$. In (11.9) $k > 2$, and for such a k

$$\begin{bmatrix}2c, 2c \\ c, c\end{bmatrix} \supset \begin{bmatrix}c + \frac{c}{k}, c + \frac{c}{k} \\ c, c\end{bmatrix}$$

so that for any $c < \min (a_1, b_1)$ and admissible in (11.11), relations (11.9) and (11.11) imply that $K \geq 1/c$. Since the c's admitted in (11.11) include values arbitrarily small, this contradiction implies that φ cannot satisfy (L_R).

12. Summary of relations between classical and MT-tests. We continue with a proof of the following theorem supplementing the theorem $D_Y \not\Longrightarrow (L_R)$ in a maximum way.

THEOREM 12.1. $J_H \not\Longrightarrow (Y_P)$.

In MT (7), Lemma 7.2 implies the existence of a function g in $\hat{F}$ over $I = \begin{bmatrix}\pi, \pi \\ 0, 0\end{bmatrix}$ with g continuous over I and vanishing on the boundary of I, and with

$$(12.1) \qquad V\begin{bmatrix}x, x \\ 0, 0\end{bmatrix} (uv\bar{g}) = \infty$$

for every $x \in (0, \pi]$. Such a function g satisfies J_H but not (Y_P) thereby establishing the theorem.

If $J_H \Longrightarrow X$ then $X \not\Longrightarrow (Y_P)$ by Th. 12.1. But we have seen in Th. 7.1 and Th. 9.1 that $J_H \Longrightarrow X$ where X is any MT-test excepting D_Y. Excepting D_Y for the moment $X \not\Longrightarrow (Y_P)$. But as seen in §10 with the aid of product functions $D_Y \not\Longrightarrow Y_P$, and since $(Y_P) \Longrightarrow Y_P$, we infer that $D_Y \not\Longrightarrow (Y_P)$.

Thus the relation

$$(12.2) \qquad X \not\Longrightarrow (Y_P)$$

holds for MT-tests X without exception.

We summarize finally all the positive implications which have been obtained. Relations of implication between MT-tests are

(a) $J_H \Longrightarrow Y \Longrightarrow Y_P \Longrightarrow L_P \Longrightarrow L_R$

(b) $J_H \Longrightarrow Y \Longrightarrow L_1 \Longrightarrow L_P \Longrightarrow L_R$

(c) $J_H \Longrightarrow V_Y \Longrightarrow L_1 \Longrightarrow L_P \Longrightarrow L_R$

(d) $D_Y \Longrightarrow V_Y \Longrightarrow L_1 \Longrightarrow L_2 \Longrightarrow L_P \Longrightarrow L_R$

obtained by combining Th. 7.1 and Th. 9.1. Relations of equivalence among MT-tests are

(e) $L_1 \equiv L_2 \equiv L_P + C^0 \equiv L_R + C^0$

(f) $L_P \equiv L_R + C^0$ (Th. 7.1) .

Implications which we have denied by producing counter-examples are

(g) $J_H \not\Longrightarrow D_Y$

(h) $Y \not\Longrightarrow V_Y$ [from (10.5)]

(i) $D_Y \not\Longrightarrow Y_P$

Other denials of implications which may be inferred from the above are indicated in Figures 1 and 2. There remains the comparison Theorem 1.1 which we now establish.

PROOF of Theorem 1.1. The first statement of Th. 1.1 is a consequence of the relation $D_Y \not\Longrightarrow (L_R)$ [Th. 11.1] taken with (d). The second statement of the theorem is implied by (12.2). That each MT-class K^X includes the corresponding classical class $K^{(X)}$ follows from the definition of the tests X and (X), recalling that

$$V(Q, g) \geq P(Q, g)$$

whenever g is defined over Q. That $K^{(X)}$ is a proper subclass of K^X is seen as follows.

<u>Case I</u>. $X \neq J_H$, Y, or Y_P. In this case $X \not\Longrightarrow (L_R)$ since $D_Y \not\Longrightarrow (L_R)$ (Th. 11.1) and $D_Y \Longrightarrow X$ by (d). Moreover $(X) \Longrightarrow (L_R)$ according to Gergen. Hence $X \not\Longrightarrow (X)$.

<u>Case II</u>. $X = J_H$, Y, or Y_P. In this case $X \not\Longrightarrow (Y_P)$ according to (12.2) while $(X) \Longrightarrow (Y_P)$ according to Gergen (1). We infer that $X \not\Longrightarrow (X)$.

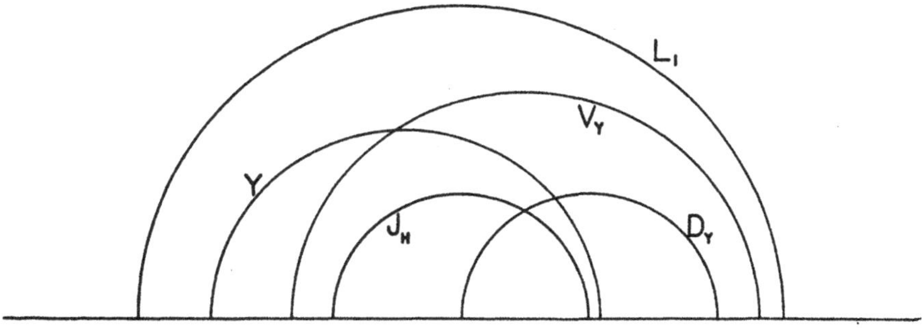

Fig. 1 MT-Tests

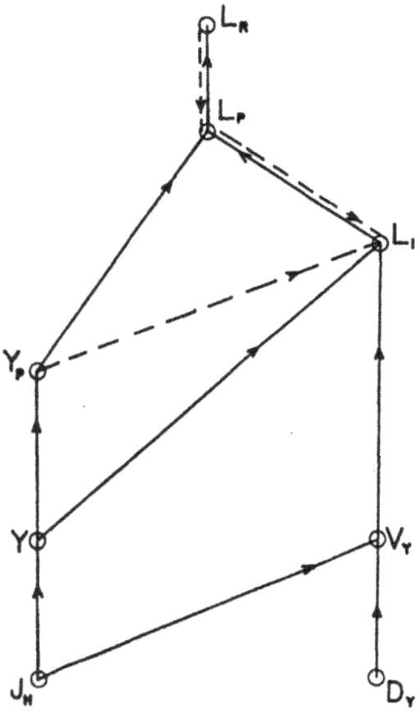

Fig. 2 MT-Tests

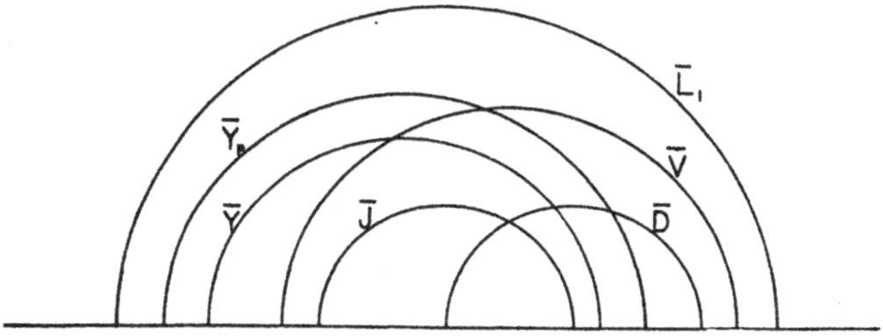

Fig. 3 One Dimensional Tests

References

S. Banach

1. "Théorie des opérations linéaires," Warsaw, 1932.

M. Fréchet

1. Sur les fonctionnelles bilinéaires, "Trans. Amer. Math. Soc.", vol. 16 (1915), pp. 215-234.

G. Fubini

1. Sulla derivata misto di un integrale doppio, II., "Rend. Circ. Mat. di Palermo," XL. (1915), pp. 297-298.

J. J. Gergen

1. Convergence criteria for double Fourier series, "Trans. Amer. Math. Soc.," vol. 35 (1933), pp. 29-63.
2. Convergence and summability criteria for Fourier series," Quarterly Jour. of Math.," Oxford Series 1 (1930), pp. 252-275.

G. H. Hardy

1. On double Fourier series, and especially those which represent the double zeta function with real and incommensurable parameters,"Quarterly Jour.," 37 (1906), pp. 53-79.
2. On certain criteria for the convergence of the Fourier series of a continuous function, "Messenger of Math.," 47 (1918), pp. 149-156.

G. H. Hardy and J. E. Littlewood

1. Young's convergence criterion for Fourier series, "Proc. London Math. Soc.," Ser 2 28(1928).

M. Morse and W. Transue

1. Functionals of bounded Fréchet variation, "Canadian Jour. Math.," vol. I (1949), pp. 153-165.
2. Functionals F bilinear over the product A x B of two pseudo-normed vector spaces, I. The representation of F., "Annals of Math.," vol. 50 (1949), pp.777-815.
3. Functionals F bilinear over the product A x B of two pseudo-normed vector spaces, II. Admissible spaces A, "Annals of Math.," vol. 51 (1950).
4. A characterization of the bilinear sums associated with the classical second variation, "Annali di Matematica Pura ed Applicata," Ser. 4 vol. 28 (1949).
5. The Fréchet variation in the small, sector limits, and left decompostions, "Canadian Jour. of Math.," (1950).
6. Integral representations of bilinear functionals, "Proc. Nat. Acad. Sci.," vol. 35 (1949), pp. 136-143.
7. Norms of distribution functions associated with bilinear functionals, In this study.
8. The Fréchet variation and the convergence of multiple Fourier series, "Proc. Nat. Acad. Sci.," 35 (1949), pp. 395-399.
9. The Fréchet variation and a generalization for multiple Fourier series of the Jordan test, "Revista di Matematica della Universita di Parma," 1(1950), pp. 1-16.
10. Implications of the Young-Pollard convergence criterion for Fourier series, "Duke Math. Jour."
11. A calculus for Fréchet variations, "Jour. Indian Math. Soc.," (1950).

References (cont.)

S. Pollard

1. On the criteria for the convergence of a Fourier series, "Journal of the London Math. Soc.," 2 (1927), pp. 255-262.

L. Tonelli

1. "Serie Trigonometriche," Bologna, 1928.

W. H. Young

1. Multiple Fourier series, "Proc. London Math. Soc.," vol. II (1913), pp. 133-184.

IV. NORMS OF DISTRIBUTION FUNCTIONS ASSOCIATED WITH BILINEAR FUNCTIONALS

By Marston Morse and William Transue[1]

0. Introduction. We are concerned with the theory of functionals bilinear on the Cartesian product $A \times B$ of two pseudo-normed vector spaces $A \times B$. The spaces A and B are linear in the sense of Banach, and are provided with pseudo-norms.[2]

The setting of the problem. Let E' be the interval $[0,1]$ of an s-axis and E" the interval $[0,1]$ of a t-axis. We shall be concerned with spaces A and B of functionals x and y respectively with values x(s) and y(t) mapping E' and E" into R_1 the axis of real numbers. Let f be a functional mapping $A \times B$ into R_1 with values f(x,y) such that[3] $f(\cdot,y)$ $[f(x,\cdot)]$ for fixed $y \in B$ $[x \in A]$ is additive and homogeneous and

$$(0.0) \qquad |f(x,y)| \leq |x|_A \, |y|_A \, M$$

for some constant M. One terms f bilinear over $A \times B$. In MT 2 five axioms are used to define admissible spaces A and B, and it has been shown that any functional f bilinear over an admissible $A \times B$ is representable by an L-S-integral (Lebesgue-Stieltjes)[4]

$$(0.1) \quad f(x,y) = \int_0^1 x(s)\, d_s \int_0^1 y(t)\, d_t k(s,t) = \int_0^1 y(s)\, d_s \int_0^1 x(t)\, d_t k(t,s),$$

---

1. The contribution of Dr. Transue to this paper is in partial fulfillment of a contract between the Office of Naval Research and the Institute for Advanced Study.

2. A pseudo-norm for an element $x \in A$ is a number $|x|_A \geq 0$ such that for x_1 and $x_2 \in A$, $|x_1+x_2|_A \leq |x_1|_A + |x_2|_A$ while for each constant c,

$$|cx|_A = |c| \, |x|_A$$

It is not required that x be the null element in A when $|x|_A = 0$.

3. The functional defined by f(x,y) for fixed y or x is denoted by $f(\cdot,y)$ or $f(x,\cdot)$ respectively.

4. First introduced as a Riemann-Stieltjes integral by Fréchet for products $C \times C$.

where k (termed a d-function) is canonical in a sense to be defined in §2 and has a finite A-B-variation h(A,B;k). This variation depends in its definition on both A and B as will be seen in §1. As the authors have shown h(A,B;k) can be characterized as the minimum of constants M for which (0.0) holds for all $(x,y) \in A \times B$. (Cf. Theorem 12.1, MT 2.)

A study of the way in which h(A,B;k) depends upon A and B is essential for an understanding of bilinear functionals f over $A \times B$. As an illustration we may refer to the problem of tests for the Pringsheim convergence of double Fourier series recently found by the authors to be intimately connected with the theory of functionals bilinear over $C \times C$. Here C is the Banach space of functionals continuous over the interval [0,1]. The test (L_R) of Gergen, p. 35, has been shown by Gergen to be not more restrictive than those earlier tests for double series which generalize the classical test of Jordan, de la Vallée Poussin, Young, Lebesgue, Pollard, etc.. By the introduction of the Fréchet variation in place of the Vitali variation in the definition of these tests we arrive at new tests, not only equally sufficient for the Pringsheim convergence of the Fourier series of φ, but also in each case definitely less restrictive in their conditions on φ. Let K_{L_R} denote the class of functions φ which satisfy our modification of Gergen's test. We show that our K_{L_R} includes as a proper (actually smaller) subclass the class of functions defined by any of the earlier tests (L_R), (L_P), (Y_P), etc. as defined by Gergen. It is by way of the functions defined in this paper that the inclusion is shown to be proper. The difficulty in such a proof may be sensed when it is recalled that this is the first proof (known to us) of a relation of proper inclusion between tests of Lebesgue type (Lebesgue, Pollard, Gergen, etc.) in the 2-dimensional case.

Admissible spaces A and B. The space of measurable functions x mapping R_1 into R_1 with $x(s) = 0$ for $s < 0$ and $s > 1$, and with a pseudo-norm given by the L-integral

$$\left[\int_0^1 |x(s)|^p \, ds\right]^{\frac{1}{p}} \qquad [p \geq 1]$$

will be denoted by L_p. The space C (Cf. Banach) is at the same time pseudo-normed in the sense of Banach. For the remainder of this paper admissible spaces A and B are to be chosen from the spaces C and L_p $(p \geq 1)$.

One object of the present paper is to show by counter example that the condition

$$(0.2) \qquad h(A,B;k) < \infty$$

on a functional k which is canonical relative to the square $E' \times E''$ (See §1), does not in general imply that the Vitali variation $V(k)$ is finite, and to compare (0.2) with the condition

$$h(A',B';k) < \infty$$

for other admissible products $A' \times B'$.

The variation $V(k)$ is that classically used in the definition of an L-S-integral. We shall show that (0.2) holds in very general cases in which $V(k) = \infty$. More definitely we shall prove the following:

(a). When $p > 1$, $q > 1$, there exists a d-function K_{pq} relative to $L_p \times L_q$ for which (0.2) holds while $V(K_{pq}) = \infty$.

(b). Statement (a) implies that there exists a d-function k relative to any one of the products

$$(0.3) \qquad C \times C, \qquad C \times L_p, \qquad L_p \times C, \qquad [p > 1]$$

such that (0.2) holds for this product while $V(k) = \infty$.

A result of the character of (a) for the product $C \times C$ is due to Clarkson and Adams (abbreviated CA) who have shown that there exists a functional k_0 with a finite Fréchet variation but with $V(k_0) = \infty$.

The product $C \times C$ is extreme in the sense that the existence of a d-function k relative to $C \times C$ with $h(C,C;k)$ finite and $V(k)$ infinite proves nothing concerning the existence of such functions for the other products in (0.3). In contrast the existence of d-functions k relative to $L_p \times L_q$ with $h(L_p,L_q;k)$ finite and $V(k)$ infinite implies the existence of similar d-functions k for each product $A \times B$ in (0.3) and for each product $L_{p_1} \times L_{q_1}$ for which $p_1 > p$, $q_1 > q$. Results of this character for $L_p \times L_q$ are thus more effective than corresponding results for $C \times C$.

A study of the product $c \times c$ has been made by Littlewood with results analogous to the above for $C \times C$. Paul Lévy has recently shown the existence of a function k which is continuous and for which $V(k) = \infty$, while the integrals (0.1) are bilinear over $L_2 \times L_2$. This possibility is also implied by our theorems. None of the earlier theorems seem to imply our results other than in these special cases.

We conclude the paper by a comparison of the Fréchet variation and Vitali variation. This comparison is relevant to order hypotheses of the character made in the theory of double Fourier series. Given $x > 0$ and $y > 0$, let $P\begin{bmatrix}x,y\\0,0\end{bmatrix}(k)$ and $V\begin{bmatrix}x,y\\0,0\end{bmatrix}(k)$ be respectively the Fréchet variation and Vitali variation of k over the rectangle with vertices (0.0) and (x,y).

Let μ be an arbitrary positive number. We consider the conditions on k relative to the product xy,

$$(0.4) \qquad P\begin{bmatrix} x,y \\ 0,0 \end{bmatrix}(k) = O[(xy)^{\mu}], \qquad [\mu > 0]$$

$$(0.5) \qquad V\begin{bmatrix} x,y \\ 0,0 \end{bmatrix}(k) = O[(xy)^{\mu}],$$

where x and y are infinitesimals. We state the following theorem.

Among functionals k which are canonical relative to $E' \times E''$ the set of functions which satisfy (0.5) is a proper subclass of the set of functions which satisfy (0.4). Moreover k can be so chosen that it is continuous over $E' \times E''$, is an indefinite integral of the form

$$k(u,v) = \int_1^u \int_1^v \varphi(s,t)\,ds\,dt, \qquad [u > 0,\ v > 0]$$

and satisfies (0.4) with

$$V\begin{bmatrix} x,y \\ 0,0 \end{bmatrix}(k) = \int_0^x \int_0^y |\varphi(s,t)|\,ds\,dt = \infty$$

for every $x > 0$, $y > 0$.

1. The spaces A_d of d-functions. This paper is concerned with counter examples and with the general principles which make it possible to construct and interpret such examples. These examples are functionals in spaces A_d now to be defined.

A function g mapping R_1 into R_1 will be termed left canonical relative to the interval $E = [0,1]$ if g has a bounded total variation T(g), if $g(s) = 0$ for $s \leq 0$, if $g(s) \equiv g(1)$ for $s \geq 1$, and if g is left continuous for $s < 1$.

A partition π of E is defined by a set of points

$$0 = s_0 < s_1 < \dots < s_{r_\pi} = 1.$$

An $\mathfrak{s}$-function η (s = "step") mapping R_1 into R_1 is said to be associated with π if $\eta(s) = 0$ for $s < 0$ and $s > 1$, and if $\eta(s)$ has a constant value η_r ($r = 1, \dots, r$) on the rth of the intervals

$$(1.1) \qquad [s_0, s_1), \ [s_1, s_2), \ \dots, \ [s_{\rho-2}, s_{\rho-1}), \ [s_{\rho-1}, s_\rho] \qquad [\rho = r_\pi].$$

The pseudo-norm $|\eta|_A$ of an s-function $\eta \in A = L_p$ is well defined. If $A = C$ we shall set

$$|\eta|_A = \max |\eta(s)|,$$

although of course η is not in C unless η is continuous. An s-function η such that $|\eta|_A = 1$ will be called a unit s-function.

The spaces A_d of d-functions.[5] Leg g be left canonical relative to E and let $I_r(g)$ be the Hellinger integral of g of index $r > 1$, taken over R_1 with respect to s. (Cf. Hildebrandt, §4.) As F. Riesz, §5, has shown, a necessary and sufficient condition that $I_r(g)$ be finite is that g be absolutely continuous, that $\dot{g}$ be in L_r, and that

$$(1.2) \qquad I_r(g) = \int_{R_1} |\dot{g}(s)|^r \, ds \qquad [r > 1].$$

Set

$$(1.3) \qquad H_r(g) = [I_r(g)]^{\frac{1}{r}}, \qquad [1 < r < \infty]$$

$$(1.4) \qquad H_\infty(g) = \sup_{s' < s''} \left| \frac{g(s'') - g(s')}{s'' - s'} \right|,$$

taking the sup over all s' and s" in R_1 with $s' < s''$. When $p > 1$, let

$$\frac{1}{p} + \frac{1}{p'} = 1 \qquad [\text{with } p' = \infty \text{ if } p = 1].$$

DEFINITION. When $A = L_p$ with $p \geq 1$, A_d is by definition the vector space of functionals g which are left canonical relative to E and for which $H_{p'}(g) < \infty$. The norm of an element g in A_d shall be

$$(1.5) \qquad |g|_{A_d} = H_{p'}(g) \qquad [1 < p' \leq \infty].$$

---

5. In a strict sense the d-functions to be defined in §1 should be called left d-functions.

This is a true norm in the sense of Banach. If $|g|_{A_d} = 0$, $g(s) \equiv 0$.

DEFINITION. When $A = C$, A_d is by definition the vector space of functionals which are left canonical relative to E and for which $T(g) < \infty$, with a norm

$$(1.6) \qquad |g|_{A_d} = T(g) .$$

It follows from the above theorem of F. Riesz that for $A = L_p$ $(p > 1)$ and $g \in A_d$,

$$(1.7) \qquad |g|_{A_d} = \left[\int_0^1 |\dot{g}(s)|^{p'} ds \right]^{\frac{1}{p'}} = L_{p'}(\dot{g}) \qquad \left[\frac{1}{p} + \frac{1}{p'} = 1 \right]$$

This has the following useful consequence. If g is in A_d with $A = L_p$ and $p > 1$, and if one sets

$$(1.8)' \qquad x(s) = c|\dot{g}(s)|^{\frac{p'}{p}} \operatorname{sign} \dot{g}(s),$$

and chooses $c > 0$ so that $|x|_{L_p} = 1$ when $|g|_{A_d} \neq 0$, then

$$(1.8)'' \qquad \int_0^1 x(s) \, \dot{g}(s) \, ds = |g|_{A_d} .$$

One establishes (1.8)" readily with the aid of (1.7), making use of the relations

$$\frac{p'}{p} + 1 = p', \qquad 1 - \frac{1}{p} = \frac{1}{p'} .$$

The spaces A admissible in this paper are admissible in the sense of MT 2, as shown in MT 3; §§ 3, 4, 5. Lemma 1.1 is then a consequence of Theorem 5.2, MT 2.

LEMMA 1.1. If g is an element in an admissible space A_d, then with the summation convention of tensor algebra,

$$(1.9)' \qquad |g|_{A_d} = \sup_{|\eta|_A = 1} |\eta_r \, \Delta_r g|, \qquad [r = 1, \ldots, r_\pi]$$

where

$$\Delta_r g = g(s_r) - g(s_{r-1}), \tag{1.9)"}$$

and where the sup is over all unit s-functions in A associated with partitions π of E.

From this lemma the following lemma can be derived.

LEMMA 1.2. *If g is an element in an admissible space A_d and is linear on the closure of each of the intervals of a fixed partition π of $[0,1]$, then (1.9)' is valid with η restricted to unit s-functions associated with the given partition*

We shall need this lemma only for the case in which $A = L_p$ $(p \geq 1)$ and will accordingly limit the proof to this case. We can suppose that $|g|_{A_d} \neq 0$ and hence $g \not\equiv 0$.

<u>The case</u> $p > 1$. In this case the lemma follows at once from (1.8)", choosing η as that unit s-function associated with π which is given by the right member of (1.8)' almost everywhere. For this choice of η, (1.8)" gives

$$\eta_r \Delta_r g = |g|_{A_d} \qquad [r = 1, \ldots, r_\pi]. \tag{1.10}$$

<u>The case</u> $p = 1$. In this case let J be a subinterval of the partition π on which $|\dot{g}(s)|$ is a maximum, and choose η as the characteristic function of J divided by the length of J. For this choice of η,

$$\eta_r \Delta_r g = \sup_{s' < s''} \left| \frac{g(s'') - g(s')}{s'' - s'} \right| = |g|_{A_d},$$

and the lemma follows.

The spaces A_d here admitted may be formally ordered by the inclusion relations

$$C_d \supset L_{pd} \supset L_{qd} \qquad [p > q \geq 1]. \tag{1.11}$$

Here L_{pd} is an abbreviation for $(L_p)_d$. We shall say that an inclusion relation is <u>proper</u> if the equality is excluded. The inclusion relations (1.11) are proper as affirmed in the following lemma.

LEMMA 1.3. In any admissible space A_d there is an element g such that g is in no space A_d' which follows A_d in (1.11).

The case A = C. A $g \in C_d$ which is not continuous is not in L_{pd}. Such a g clearly exists so that the lemma is true for A = C.

The case $A = L_p$. For fixed $p > 1$, any $g \in L_{pd}$ is of the form

$$g(s) = \int_0^s y(u)\, du, \tag{1.12}$$

where y is in $L_{p'}$, with $y(u) = 0$ for $u < 0$ and $u > 1$. For $p > q > 1$ one has $q' > p' > 1$. For fixed $p > 1$ one can readily construct a $y \in L_{p'}$ with $y \notin L_{q'}$ and $y(u) = 0$ for $u < 0$ and $u > 1$. For such a y, g as given by (1.12) is in L_{pd}, but not in L_{qd} by Riesz's theorem. [Cf. (1.2)]. Further, such a g cannot be in L_d. For in that case $|y|$ would be bounded almost everywhere and g would be in every space L_{qd}, contrary to the choice of g.

For spaces A admissible in the sense of this paper (or admissible in the sense of MT 2, § 3),

$$|g_1 + g_2|_{A_d} \leq |g_1|_{A_d} + |g_2|_{A_d} \qquad [r > 1]. \tag{1.13}$$

The following lemma extends (1.13).

LEMMA 1.4. For each integer $n > 0$, let g_n map R_1 into R_1 and suppose that

$$g_1(s) + g_2(s) + \dots \tag{1.14}$$

converges to g(s) for each s. Then

$$|g|_{A_d} \leq |g_1|_{A_d} + |g_2|_{A_d} + \dots \qquad [r > 1]. \tag{1.15}$$

Relation (1.15) is trivial if the series in (1.15) diverges or if $|g_n|_{A_d} = \infty$ for some n. Suppose then that the series on the right of (1.15) converges to a finite sum. Set

$$S_n(s) = g_1(s) + g_2(s) + \dots + g_n(s).$$

By virtue of the lower semi-continuity of $|g|_{A_d}$ (Cf. Theorem 5.3, MT 2),

(1.16)
$$|g|_{A_d} \leq \liminf_{n\to\infty} |S_n|_{A_d} .$$

But

(1.17)
$$|S_n|_{A_d} \leq |g_1|_{A_d} + |g_2|_{A_d} + \cdots + |g_n|_{A_d}$$

in accordance with (1.13). Relation (1.15) follows from (1.16) and (1.17).

A fundamental lemma concerns a transformation

(1.18)
$$u = as + b, \qquad [0 < b < a + b < 1]$$

which maps $E = [0,1]$ into a subinterval $[b,a+b]$ of E.

LEMMA 1.5. Let g map R_1 into R_1 and be left canonical relative to E. Under the transformation (1.18) set $g(s) = \bar{g}(u)$. Then $\bar{g}$ is left canonical relative to E and

(1.19)
$$|\bar{g}|_{L_{pd}} = a^{-\frac{1}{p}} |g|_{L_{pd}} .$$

To establish (1.19) recall that

$$|g|_{L_{pd}} = [I_{p'}(g)]^{\frac{1}{p'}}, \qquad [\tfrac{1}{p} + \tfrac{1}{p'} = 1]$$

when $p > 1$, by definition of $|g|_{L_{pd}}$. By definition of the Hellinger integral

(1.20)
$$I_{p'}(g) = \sup_{\pi} \frac{|\Delta_r g|^{p'}}{|\Delta_r s|^{p'-1}}, \qquad [r \text{ summed}]$$

taking the sup over all partitions π of $[0,1]$. By virtue of (1.18), $\Delta u = a\,\Delta s$ so that

(1.21)
$$I_{p'}(\bar{g}) = a^{1-p'} I_{p'}(g).$$

On taking the p'th root of both members of (1.21) and using the relation $(1-p')/p' = -1/p$, (1.19) follows when $p > 1$. When $p = 1$,

$$|g|_{L_d} = \sup_{s' < s''} \left| \frac{\Delta g}{\Delta s} \right| \qquad [\Delta s = s'' - s']$$

and (1.19) follows from the relation $\Delta u = a \Delta s$.

2. <u>The variation</u> $h(A,B;k)$. Let R_2 be the Cartesian space of coordinates (s,t). Let E' and E" be respectively the unit interval [0,1] on the s and t axes of R_2. A function k mapping R_2 into R_1 will be said to be <u>left canonical</u> relative to E' $\times$ E" if $k(\cdot,t)$, $[k(s,\cdot)]$ is left canonical in the sense of §2 relative to E' [E"] for each fixed $t \in R_1$ $[s \in R_1]$.

A partition π of E' $\times$ E" is defined by a partition π' of E' of the form (1.1), and a second independent partition π'' of E",

$$(2.1) \qquad 0 \leq t_0 < t_1 < \dots < t_{n_\pi} = 1.$$

Let η and ζ be s-functions associated respectively with π' and π''. Let k be canonical relative to E' $\times$ E". Set

$$\Delta_{rn}(k) = k(s_r,t_n) - k(s_r,t_{n-1}) - k(s_{r-1},t_n) + k(s_{r-1},t_{n-1})$$

for each $r = 1, \dots, r_\pi$, $n = 1, \dots, n_\pi$. Employing the summation convention of tensor algebra, as always, set

$$(2.2) \qquad \sup_{\eta,\zeta} \eta_r \zeta_n \Delta_{rn}(k) = h(A,B;k),$$

taking the sup over all partitions π of E' $\times$ E" and associated s-functions η, ζ for which

$$|\eta|_A = 1, \qquad |\zeta|_B = 1.$$

We term $h(A,B;k)$ the A-B-variation of k, admitting that this variation may be infinite. A function k which is canonical relative to E' $\times$ E" and for which $h(A,B;k) < \infty$ will be termed a d-<u>function</u> relative to A $\times$ B.

A function $x \in A$ for which $|x|_A = 1$ is termed <u>unit</u> in A, and if $|x|_A \leq 1$, is termed <u>subunit</u> in A. In defining $h(A,B;k)$ it is immaterial whether the sup in (2.2) is taken over unit elements in A and B respectively or over subunit elements.

As stated in §2, the spaces A and B are to be chosen from the spaces

C and L_p $(p \geq 1)$. It follows from MT 3, §§ 3, 4, 5, that these spaces satisfy Conditions I - V of MT 2. To give another general property let functions

$$\Delta_r^{(1)}(k) = k(s_r,\cdot) - k(s_{r-1},\cdot), \qquad [r = 1, \ldots, r_\pi]$$

$$\Delta_n^{(2)}(k) = k(\cdot,t_n) - k(\cdot,t_{n-1}), \qquad [n = 1, \ldots, n_\pi]$$

mapping R_1 into R_1 be introduced, corresponding respectively to partitions π' and π'' of E' and E". According to Theorem 7.1 of MT 2,

$$h(A,B;k) = \sup_{|\eta|_A = 1} |\eta_r \Delta_r^{(1)}(k)|_{B_d}, \tag{2.3}$$

$$h(A,B;k) = \sup_{|\xi|_B = 1} |\xi_n \Delta_n^{(2)}(k)|_{A_d}, \tag{2.4}$$

where η and ξ are s-functions associated respectively with π' and π''.

The following lemma is an extension of Lemma 1.2.

LEMMA 2.1. Let π' and π'' be fixed partitions of E' and E" respectively. Let k be left canonical relative to E' $\times$ E" with $k(\cdot,t)$ linear on the closure of each interval of π' for each fixed $t \in R_1$, and $k(s,\cdot)$ linear on the closure of each interval of π'' for each fixed $s \in R_1$. Then for $p \geq 1$, $q \geq 1$,

$$h(L_p,L_q;k) = \sup_{\eta,\xi} \eta_r \xi_n \Delta_{rn}(k), \tag{2.5}$$

taking the sup over all subunit s-functions η and ξ in L_p and L_q respectively, restricting η and ξ to s-functions associated with the given partitions π' and π'' respectively.

Let π be an arbitrary partition of E' $\times$ E" and $(\bar{\eta}, \bar{\xi})$ an associated pair of s-functions subunit in L_p and L_q respectively. With a sup taken over all such $(\bar{\eta}, \bar{\xi})$, $h(L_p,L_q;k)$ equals

$$\sup_{\bar{\eta}} \left[\sup_{\bar{\xi}} (\bar{\xi}_n \Delta_n^{(2)} [\bar{\eta}_r \Delta_r^{(1)} (k)]) \right].$$

But for fixed $\bar{\eta}$, $\bar{\eta}_r \Delta_r [k(\cdot,t)]$ is linear in t over the closure of each interval of π''. It follows from Lemma 1.2 that for fixed $\bar{\eta}$,

$$\sup_{\xi} (\bar{\xi}_n \overset{(2)}{\Delta}_n [\bar{\eta}_r \overset{(1)}{\Delta}_r (k)]) = \sup_{\xi} (\xi_n \overset{(2)}{\Delta}_n [\bar{\eta}_r \overset{(1)}{\Delta}_r (k)]) .$$

restricting ξ as in the lemma. Hence $h(L_p,L_q;k)$ equals

$$\sup_{\xi} \left[\sup_{\bar{\eta}} (\bar{\eta}_r \overset{(1)}{\Delta}_r [\xi_n \overset{(2)}{\Delta}_n (k)]) \right],$$

restricting ξ as in the lemma and taking $\bar{\eta}$ as above. It follows by another application of Lemma 1.2 that the $\bar{\eta}$ can be replaced by the s-functions η of the lemma so that (2.5) holds as stated.

The following theorem is a consequence of Lemma 7.1 and Theorem 7.1 of MT 2. One must, however, note that when A and B are the spaces L_p or C, one can take $M_A(s) = M_B(s) \equiv 1$ in Lemma 7.1 of MT 2.

THEOREM 2.1. If k is left canonical relative to $E' \times E''$, then for fixed $s \in E'$ and $t \in E''$,

$$(2.6)' \qquad |k(\cdot,t)|_{A_d} \leqq h(A,B;k),$$

$$(2.6)'' \qquad |k(s,\cdot)|_{B_d} \leqq h(A,B;k).$$

Relations (2.6) imply in particular that

$$(2.7)' \qquad T[k(\cdot,t)] \leqq h(C,B;k),$$

$$(2.7)'' \qquad T[k(s,\cdot)] \leqq h(A,C;k),$$

$$(2.8)' \qquad H_{p'}[k(\cdot,t)] \leqq h(L_p,B;k), \qquad [1 < p' \leqq \infty]$$

$$(2.8)'' \qquad H_{p'}[k(s,\cdot)] \leqq h(A,L_p;k),$$

for arbitrary admissible A and B. For the definition of $H_{p'}$ see (1.3) and (1.4).

The implications of these relations will be illustrated in the case of (2.8). If $h(L_p,B;k)$ is finite, then (2.8)' implies that $k(\cdot,t)$ is absolutely continuous for each fixed t and is the integral of a function in $L_{p'}$

[Cf. (1.2)]. If $h(L,B;k)$ is finite, then (2.8)' implies that $k(\cdot,t)$ satisfies a Lipschitz condition

$$| k(s'',t) - k(s',t) | \leq h(L,B;k) |s'' - s'|,$$

in accordance with the definition of H_∞. Relations (2.7) in the case in which $A = B = C$ were established by Fréchet.

We conclude this section with two technical lemmas which follow from Lemmas 1.4 and 1.5 respectively. Lemmas 2.2 and 2.3 will be used in §6.

LEMMA 2.2. *For each integer $n > 0$, let k_n be a function mapping R_2 into R_1, left canonical relative to $E' \times E''$ and such that the series*

$$k_1(s,t) + k_2(s,t) + \dots = k(s,t)$$

converges for each (s,t) in R_2 and defines a function k which is left canonical relative to $E' \times E''$. Then for admissible A and B,

$$h(A,B;k) \leq h(A,B;k_1) + h(A,B;k_2) + \dots . \tag{2.9}$$

Let $\mathfrak{z}$ be an arbitrary s-function associated with a partition of E". Then for a fixed s-function $\mathfrak{z}$,

$$|\mathfrak{z}_n \overset{(2)}{\Delta}_n (k)|_{A_d} \leq |\mathfrak{z}_n \overset{(2)}{\Delta}_n (k_1)|_{A_d} + |\mathfrak{z}_n \overset{(2)}{\Delta}_n (k_2)|_{A_d} + \dots \tag{2.10}$$

in accordance with Lemma 1.4. Relation (2.9) results on taking the sup of the respective members of (2.10) over all unit s-functions $\mathfrak{z}$ in B, using (2.4).

LEMMA 2.3. *Let k be a d-function relative to $L_p \times L_q$ for $p > 1$, $q > 1$. If under a transformation*

$$\begin{aligned} u &= as + b, \\ v &= at + b, \end{aligned} \qquad 0 < b < a+b < 1 \tag{2.11}$$

$\bar{k}(u,v) = k(s,t)$, then $\bar{k}$ is a d-function relative to $E' \times E''$ for which

$$h(L_p,L_q;\bar{k}) = a^{-\frac{1}{p}\frac{1}{q}} h(L_p,L_q;k). \tag{2.12}$$

We begin by making the transformation

$$(2.13) \qquad u = as + b, \qquad t = t,$$

under which $k(s,t) = \kappa(u,t)$. We shall apply (1.19), recalling that

$$(2.14)' \qquad h(L_p, L_q; k) = \sup_{\mathfrak{f}} |\mathfrak{f}_n \Delta_n^{(2)}(k)|_{L_{pd}},$$

$$(2.14)'' \qquad h(L_p, L_q; \kappa) = \sup_{\mathfrak{f}} |\mathfrak{f}_n \Delta_n^{(2)}(\kappa)|_{L_{pd}},$$

and taking the sup over all unit s-functions $\mathfrak{f}$ in L_q associated with partitions of E". Set

$$\mathfrak{f}_n \Delta_n^{(2)}(k) = g,$$

$$\mathfrak{f}_n \Delta_n^{(2)}(\kappa) = \bar{g},$$

and observe that $g(s) = \bar{g}(u)$ under (2.13). It then follows from (2.14) and (1.19) applied to g and $\bar{g}$ that

$$(2.15) \qquad h(L_p, L_q; \kappa) = a^{-\frac{1}{p}} h(L_p, L_q; k).$$

If the transformation (2.3) is followed by the transformation $u = u$, $v = at + b$, so that $\bar{k}(u,v) = \kappa(u,t) = k(s,t)$, one finds that

$$(2.16) \qquad h(L_p, L_q; \bar{k}) = a^{-\frac{1}{q}} h(L_p, L_q; \kappa),$$

and (2.12) follows from (2.15) and (2.16).

3. Preliminary comparisons. We begin with a general Comparison Theorem.

THEOREM 3.0. If k is a d-function relative to any admissible product $A' \times B'$, and if $A \times B$ is a second admissible product such that

$$(3.1) \qquad A \subset A', \qquad B \subset B',$$

then

$$(3.2) \qquad h(A,B;k) \leq h(A',B';k).$$

The condition $A \subset A'$ $[B \subset B']$ implies the relation

$$|\eta|_A \geq |\eta|_{A'}, \qquad [|\xi|_B \geq |\xi|_{B'}]$$

for every s-function η $[\xi]$ associated with a partition of E' [E"] respectively. (For example $|\eta|_C \geq |\eta|_L$ since $C \subset L$.) Hence

$$(3.3) \qquad \sup_{\eta,\xi} \eta_r \xi_n \Delta_{rn}(k),$$

taken over the unit s-functions η and ξ in A and B respectively, equals a sup of the form (3.3) taken over a subclass of subunit η and ξ in A' and B' respectively. Thus (3.2) holds.

<u>Product</u> d-<u>functions</u> $k = \varphi\psi$. General counterexamples of a simple type can be given in terms of d-functions k of the form

$$(3.4) \qquad k(s,t) = \varphi(s)\,\psi(t).$$

On the other hand we shall see that such product d-functions do not include d-functions with finite Fréchet variation and infinite Vitali variation. We are thereby compelled in §§4, 5, 6 to construct d-functions of a more complicated type.

If the function k given by (3.4) does not vanish identically, a necessary and sufficient condition that $\varphi\psi$ be left canonical relative to $E' \times E"$ is that φ and ψ be left canonical relative to E' and E" respectively. This is an immediate consequence of the definitions involved. If $\varphi\psi \neq 0$, and if $\varphi\psi$ is left canonical relative to $E' \times E"$, then

$$(3.5) \qquad h(A,B;\varphi\psi) = |\varphi|_{A_d}\,|\psi|_{B_d},$$

as one verifies immediately. [Cf. (1.9)'].

Recall that the condition $|g|_{A_d} = 0$ on a g which is left canonical relative to E' is equivalent to the condition $g(s) = 0$ for every s. It follows from (3.5) that a necessary and sufficient condition that a product $\varphi\psi$ be a d-function relative to $A \times B$ is that φ be in A_d and ψ be in B_d, provided $\varphi\psi \neq 0$, that is, provided neither $|\varphi|_{A_d}$ nor $|\psi|_{B_d} = 0$.

Theorem 3.1 follows from Lemma 1.3 on making use of the product d-function $\varphi\psi$.

THEOREM 3.1. *There exists a d-function $\varphi\psi$ relative to any admissible product $A \times B$, such that $\varphi\psi$ is not a d-function relative to any admissible product $A' \times B'$ for which*

$$A \subset A', \qquad B \subset B',$$

and for which at least one of these inclusions is proper.

In accordance with Lemma 1.3, there exists a φ in A_d which is not in any A'_d which follows A_d in (1.11). Similarly there exists a ψ in B_d which is not in any B'_d which follows B_d in (1.11). By hypothesis, if $A' = A$ $[B' = B]$, then $B \neq B'$ $[A \neq A']$. If $A' \neq A$, A'_d follows A_d in (1.11), and if $B' \neq B$, B'_d follows B_d in (1.11). If $\varphi\psi$ were a d-function relative to $A' \times B'$, φ would be in A'_d and ψ in B'_d according to the conclusion just prior to the theorem, and contrary to the choice of φ and ψ.

The Vitali variation. Let k be left canonical relative to $E' \times E''$. The Vitali variation of k is then

$$V(k) = \sup_{\pi} \sum_{r,n} |\Delta_{rn}(k)|, \qquad [r = 1,\ldots,r_{\pi},\ n = 1,\ldots,n_{\pi}]$$

taking the sup over all partitions π of $E' \times E''$.

THEOREM 3.2. *For any k which is left canonical relative to $E' \times E''$,*

$$h(C,C;k) \leqq V(k). \tag{3.6}$$

There exist functions k which are left canonical relative to $E' \times E''$ for which $0 < h(C,C;k) < \infty$, while $V(k) = \infty$.

Relation (3.6) is an immediate consequence of the definitions involved. For any one of the d-functions K_{pq} defined in §6,

$$V(K_{pq}) = \infty, \qquad 0 < h(C,C;K_{pq}) < \infty.$$

For admissible products $A \times B$ other than $C \times C$ no relation analogous to (3.6) holds in general. The following theorem throws light on this situation.

THEOREM 3.3. If φ and ψ are left canonical relative to $[0,1]$, and if $\varphi\psi \not\equiv 0$, then for admissible spaces A and B,

$$(3.7) \quad V(\varphi\psi) = T(\varphi)T(\psi) \leqq |\varphi|_{A_d} |\psi|_{B_d} = h(A,B;\varphi\psi).$$

If $A \times B$ is not the product $C \times C$, there exists a product left canonical relative to $E' \times E''$ such that

$$(3.8) \qquad V(\varphi\psi) < \infty, \qquad h(A,B;\varphi\psi) = \infty.$$

The first equality in (3.7) is trivial. The end equality in (3.7) is implied by (3.5). The middle relation is a consequence of (3.2), since $T(\varphi)T(\psi) = h(C,C;\varphi\psi)$ and

$$C \subset A, \qquad C \subset B.$$

If $A \times B$ is not the product $C \times C$, then either A, or else B, say A, properly includes C. It follows from Lemma 1.3 that C_d properly includes A_d. Let φ then be in C_d but not in A_d, and let ψ be in B_d with $\psi \not\equiv 0$. Then $T(\varphi)$ and $T(\psi)$ are finite so that $V(\varphi\psi) < \infty$. Moreover $|\varphi|_{A_d} = \infty$ since φ is left canonical relative to E but not in A_d; and $|\psi|_{B_d} > 0$ since $\psi \not\equiv 0$. Hence

$$h(A,B;\varphi\psi) = |\varphi|_{A_d} |\psi|_{B_d} = \infty.$$

Thus (3.8) holds.

The relation between $V(k)$ and $h(A,B;k)$ will be completed for our immediate purposes on proving the following theorem in §6.

THEOREM 3.4. Relative to any product $A \times B$ chosen arbitrarily from the products,

$$C \times C, \qquad C \times L_p, \qquad L_p \times C, \qquad L_p \times L_q, \qquad [p > 1,\ q > 1]$$

there exists a d-function K for which $V(K) = \infty$.

Since $h(A,B;K) < \infty$ for any d-function K satisfying the theorem, K cannot be of the form $\varphi\psi$ without violating the relation

$$V(\varphi\psi) \leqq h(A,B;\varphi\psi)$$

in (3.7). When $A \times B = L_p \times L_q$ with $p > 1$, $q > 1$, K is astonishingly regular even although $V(K) = \infty$. In fact $K(\cdot,t)$ and $K(s,\cdot)$ are absolutely continuous for fixed t and s respectively, with $K_s(\cdot,t)$ in L_p, and $K_t(s,\cdot)$ in L_q, in accordance with (2.8). Moreover it follows from the representation

$$F(z^s_{E'}, z^t_{E''}) = K(s,t) \qquad [F = F^e_{12}]$$

of K given in Corollary 12.1 of MT 2, that K is continuous over R_2 by virtue of the continuity of F over $L_p \times L_q$ (F is an L-S-integral of the form (0.1) with $k = K$). We shall make no use of this fact.

To establish Theorem 3.4 it is necessary to make a deep preliminary study of finite bilinear functionals associated with finite matrices of elements ± 1. This is the object of the next section.

4. The matrices **a** and **c**. In this section we shall derive certain properties of a matrix (Cf. CA, §4)

$$\mathbf{a} = \| a_{rn} \| \qquad [n = 1, \ldots, m; \; r = 1, \ldots, \omega]$$

of constants $a_{rn} = \pm 1$, and then specialize **a** as **c**. We refer to the set of elements $a_{r1}, \ldots, a_{rn}$ as the rth column of **a**, and the set of elements $a_{1n}, \ldots, a_{\omega n}$ as the nth row of **a**. This notation is adapted to the use of these matrices in defining d-functions.

Let R_ω and R_m respectively denote the spaces of vectors

$$\alpha = (\alpha_1, \ldots, \alpha_\omega), \qquad \beta = (\beta_1, \ldots, \beta_m).$$

The vector α may be regarded as defining a vector

$$x = (\alpha_1, \ldots, \alpha_\omega, 0, 0, \ldots)$$

in any one of the Banach spaces l_p ($p \geq 1$) or c, with its norm $|x|_{l_p}$ or $|x|_c$. With x so given we set

$$(4.0) \qquad |x|_{l_p} = |\alpha|_p ; \qquad |x|_c = |\alpha|_c .$$

Norms for vectors β in R_m are similarly defined. With this understood, set

$$(4.1) \qquad F_{pq}(\mathfrak{a}) = \sup_{|\alpha|_p = |\beta|_q = 1} \alpha_r \beta_n a_{rn}, \qquad [r = 1, \ldots, \omega;\ n = 1, \ldots, m]$$

taking the sup over all vectors $\alpha \in R_\omega$ and $\beta \in R_m$ for which $|\alpha|_p = |\beta|_q = 1$. the following lemma gives analogues of (2.3) and (2.4). As previously,

$$\frac{1}{p} + \frac{1}{p'} = 1, \qquad \frac{1}{q} + \frac{1}{q'} = 1.$$

LEMMA 4.1. For $\alpha \in R_\omega$ and $\beta \in R_m$,

$$(4.2)' \qquad [F_{pq}(\mathfrak{a})]^{p'} = \sup_{|\beta|_q = 1} \sum_r |\beta_n a_{rn}|^{p'}, \qquad [p > 1,\ q \geq 1]$$

$$(4.2)'' \qquad F_{1q}(\mathfrak{a}) = \sup_{|\beta|_q = 1} [\max_r |\beta_n a_{rn}|], \qquad [q \geq 1]$$

$$(4.3)' \qquad [F_{pq}(\mathfrak{a})]^{q'} = \sup_{|\alpha|_p = 1} \sum_n |\alpha_r a_{rn}|^{q'}, \qquad [p \geq 1,\ q > 1]$$

$$(4.3)'' \qquad F_{p1}(\mathfrak{a}) = \sup_{|\alpha|_p = 1} [\max_n |\alpha_r a_{rn}|], \qquad [p \geq 1].$$

PROOF. We start with the relation

$$\sup_{|\alpha|_p = |\beta|_q = 1} \alpha_r \beta_n a_{rn} = \sup_{|\beta|_q = 1} [\sup_{|\alpha|_p = 1} \alpha_r b_r(\beta)],$$

where $b(\beta)$ is a vector in R_ω with components

$$b_r(\beta) = \beta_n a_{rn}, \qquad [r = 1, \ldots, \omega].$$

According to the algebra of vectors in R_ω,

$$\sup_{|\alpha|_p = 1} \alpha_r b_r(\beta) = |b(\beta)|_{p'}. \qquad [p > 1]$$

$$\sup_{|\alpha|_1 = 1} \alpha_r b_r(\beta) = \max_r |b_r(\beta)|, \qquad [p = 1].$$

Equations (4.2) follow. The proof of (4.3) is similar.

For any β in R_m and for $p > 1$, $q > 1$, set

$$\sum_r |\beta_n a_{rn}|^{p'} = \varphi_p(\beta). \tag{4.4}$$

The value of $[F_{pq}(a)]^{p'}$ is the maximum of φ_p subject to the condition $|\beta|_q = 1$. Such a maximum clearly exists. We shall establish the following lemma.

LEMMA 4.2. For any vector β in R_m which maximizes φ_p subject to the condition $|\beta|_q = 1$,

$$[F_{pq}(a)]^{p'} \leq m^{\frac{1}{q'}} \sum_r |\beta_n a_{rn}|^{p'-1} \qquad [p > 1,\ q > 1]. \tag{4.5}$$

Write

$$\sum_r |\beta_n a_{rn}|^{p'} = \sum_r |\beta_n a_{rn}|^{p'-1} |\beta_n a_{rn}| . \tag{4.6}$$

Subject to the condition $|\beta|_q = 1$,

$$|\beta_n a_{rn}| \leq \left[\sum_n |a_{rn}|^{q'}\right]^{\frac{1}{q'}} = m^{\frac{1}{q'}}, \tag{4.7}$$

so that when $|\beta|_q = 1$,

$$\sum_r |\beta_n a_{rn}|^{p'} \leq m^{\frac{1}{q'}} \sum_r |\beta_n a_{rn}|^{p'-1}. \tag{4.8}$$

Lemma 4.2 follows from (4.8) and (4.2)'.

For the purposes of the following lemma set

$$F_{pq}(a) = \bar{F}_{p'q'}(a) \qquad [p > 1,\ q > 1]. \tag{4.9}$$

LEMMA 4.3. For $2 > p > 1$ and $q > 1$,

$$[\bar{F}_{p'q'}(a)]^{p'} \leq m^{\frac{1}{q'}} [\bar{F}_{(p'-1)q'}(a)]^{p'-1} \tag{4.10}$$

This is an immediate consequence of (4.5) and (4.2)', applying (4.2)' to show that

$$[\bar{F}_{(p'-1)q'}(a)]^{p'-1} = \sup_{|\beta|_q = 1} \sum_r |\beta_n a_{rn}|^{p'-1}. \tag{4.11}$$

This application of (4.2)' is permissible if the conjugate of $p'-1$ exceeds 1, and this is the case if $p' > 2$, that is, if $2 > p$.

LEMMA 4.4. $F_{pq}(a)$ varies continuously with (p,q) for fixed (a) and $p \geq 1$, $q \geq 1$.

Given $0 < \mu < p$, $0 < \nu < q$, it is well known (Cf. HLP,[6] 2.9.1 and 2.10.3) that

$$|\alpha|_p \leq |\alpha|_\mu \leq \omega^{\frac{1}{\mu} - \frac{1}{p}} |\alpha|_p, \tag{4.12$'$}$$

$$|\beta|_q \leq |\beta|_\nu \leq m^{\frac{1}{\nu} - \frac{1}{q}} |\beta|_q. \tag{4.12$''$}$$

From the definition of $F_{pq}(a)$ it is clear that for α in R_ω and β in R_m,

$$|\alpha_r \beta_n a_{rn}| \leq |\alpha|_\mu |\beta|_\nu F_{\mu\nu}(a), \quad [\mu,\nu \text{ not summed}] \tag{4.13$'$}$$

$$|\alpha_r \beta_n a_{rn}| \leq |\alpha|_p |\beta|_q F_{pq}(a), \quad [p,q \text{ not summed}] \tag{4.13$''$}$$

We have

$$F_{\mu\nu}(a) = \sup_{|\alpha|_\mu = |\beta|_\nu = 1} |\alpha_r \beta_n a_{rn}| \leq F_{pq}(a), \tag{4.14$'$}$$

6. Hardy, Littlewood, Polya.

using (4.13)" and (4.12). Similarly,

$$(4.14)" \quad F_{pq}(\mathbf{a}) = \sup_{|\alpha|_p = |\beta|_q = 1} |\alpha_r \beta_n a_{rn}| \leqq \omega^{\frac{1}{\mu}-\frac{1}{p}} m^{\frac{1}{\nu}-\frac{1}{q}} F_{\mu\nu}(\mathbf{a}),$$

using (4.13)' and (4.12). Thus

$$(4.15) \quad F_{\mu\nu}(a) \leqq F_{pq}(a) \leqq \omega^{\frac{1}{\mu}-\frac{1}{p}} m^{\frac{1}{\nu}-\frac{1}{q}} F_{\mu\nu}(a).$$

Given $e > 0$ and fixed μ, ν, the extreme members of (4.15) have an absolute difference less than e if $|\mu - p|$ and $|q - \nu|$ are sufficiently small. The continuity of $F_{pq}(\mathbf{a})$ at the point (μ, ν) is a consequence.

With Clarkson and Adams, p. 840, we now take m as an odd integer > 1 and $\omega = 2^{m-1}$. The elements c_{r1} $(r = 1, \ldots, \omega)$ of the first row of $\mathbf{a}$ shall be 1, and the set of the remaining elements of the respective ω columns of $\mathbf{a}$ shall be taken as the ω different permutations of m-1 elements, each of which is +1 or -1. Denote the resulting matrix $\| c_{rn} \|$ by $\mathbf{c}$. From this point on we depart from Clarkson and Adams.

Set

$$F_{pq}(\mathbf{c}) = F_{pq}^{(m)} \qquad [p \geqq 1, \ q \geqq 1].$$

We shall evaluate $F_{pq}^{(m)}$ for an infinite sequence of pairs (p_n, q_n) tending to $(1,1)$ as $n \to \infty$ with $p_n > 1$, $q_n > 1$. This evaluation will lead to a proof of Theorem 5.1. A sequence of lemmas is needed.

LEMMA 4.5. $[F_{22}^{(m)}]^2 = \omega, \quad (\omega = 2^{m-1}).$

This is a consequence of the formula [Cf. (4.4)],

$$(4.16) \quad \varphi_2(\beta) = \sum_r |\beta_n c_{rn}|^2 = \omega(\beta_1^2 + \ldots + \beta_m^2),$$

which we shall now verify. Observe that in the quadratic form φ_2 the coefficient of β_n^2 is $\sum_r c_{rn}^2 = \omega$. That the coefficient $2c_{rs}c_{rn}$ of $\beta_s \beta_n$ vanishes when $n \neq s$ may be seen as follows. Suppose first that neither n nor $s = 1$. The pair (c_{rs}, c_{rn}) is one of the following pairs,

$$(1, 1), \qquad (1, -1), \qquad (-1, 1), \qquad (-1, -1),$$

and occurs in the rth column of $\mathbf{c}$. By virtue of the construction of $\mathbf{c}$, for fixed s and n there are as many columns in which this pair is (1,1) as columns in which this pair is (1, -1). Similarly, there are as many columns in which the pair (c_{rs}, c_{rn}) is (-1, -1) as (-1, 1). Hence the sum $2c_{rs}c_{rn} = 0$ when $n \neq s$ and $n > 1$, $s > 1$. When $s = 1$ and $n > 1$, there are as many columns in which $(c_{r1}, c_{rn}) = (1, 1)$ as columns in which it equals (1, -1), so that the sum $c_{r1}c_{rn} = 0$ again.

The lemma follows, since $[F_{22}^{(m)}]^2$ is the maximum of $\varphi_2(\beta)$ subject to the condition $|\beta|_2 = 1$.

LEMMA 4.6. For $1 < q \leq 2$, $[F_{2q}^{(m)}]^2 = \omega$.

According to (4,2)' and (4.16), for $1 < q \leq 2$,

$$(4.17) \qquad [F_{2q}^{(m)}]^2 = \sup_{|\beta|_q = 1} \sum_r |\beta_n c_{rn}|^2 = \omega \sup_{|\beta|_q = 1} [|\beta|_2]^2 \leq \omega.$$

The equality prevails since $\sup [|\beta|_2]^2 = 1$ for $1 < q \leq 2$, $|\beta|_q = 1$. The equality holds in particular on taking all of the components of β, as 0 except $\beta_1 = 1$.

LEMMA 4.7. For $1 < q \leq 2$ and $p' = 2, 3, 4, \ldots$,

$$(4.18) \qquad [F_{pq}^{(m)}]^{p'} \leq \omega\, m^{\frac{p'-2}{q'}}.$$

Relation (4.18) holds when $p = p' = 2$ in accordance with Lemma 4.6. Proceeding inductively, we shall assume that (4.18) holds when $p' > 2$ is replaced by $p' - 1$ and p replaced by the conjugate of $p'-1$. By Lemma 4.3,

$$[\bar{F}_{p'q'}(\mathbf{c})]^{p'} \leq m^{\frac{1}{q'}} [\bar{F}_{(p'-1)q'}(\mathbf{c})]^{p'-1},$$

and by our inductive hypothesis this becomes

$$\leq m^{\frac{1}{q'}} \omega m^{\frac{p'-3}{q'}} = \omega m^{\frac{p'-2}{q'}},$$

establishing (4.18).

The exact evaluations in the next lemma are suggested by the formula [derived from (4.18)],

$$F_{pq}^{(m)} \leqq \omega^{\frac{1}{p'}} m^{\frac{1}{q'} - \frac{2}{p'q'}}, \qquad [p > 1,\ q > 1]$$

on letting $p \to 1$ or $q \to 1$ and making use of the continuity of $F_{pq}^{(m)}$ with respect to p and q, as affirmed in Lemma 4.4.

LEMMA 4.8. $F_{11}^{(m)} = 1$, while for $p > 1$, $q > 1$,

$$(4.19) \qquad F_{p1}^{(m)} = \omega^{\frac{1}{p'}}, \qquad F_{1q}^{(m)} = m^{\frac{1}{q'}}.$$

The following direct proof is given. For $p \geqq 1$, (4.3)" gives

$$(4.20) \qquad F_{p1}^{(m)} = \sup_{|\alpha|_p = 1} [\max_n |\alpha_r c_{rn}|] = \max_n [\sup_{|\alpha|_p = 1} |\alpha_r c_{rn}|],$$

so that for $p > 1$,

$$F_{p1}^{(m)} = \max_n [|c_{1n}|^{p'} + \dots + |c_{\omega n}|^{p'}]^{\frac{1}{p'}} = \omega^{\frac{1}{p'}},$$

For $p = 1$, again using (4.3)",

$$(4.21) \qquad F_{11}^{(m)} = \max_n [\max_r |c_{rn}|] = 1.$$

The evaluation of $F_{1q}^{(m)}$ is similar to that of $F_{p1}^{(m)}$, interchanging the roles of α and β.

A result (4.24) of Clarkson and Adams, p. 840, is easily verified. With reference to the Banach space c, set

$$(4.22) \qquad F^{(m)} = \sup_{|\alpha|_c = |\beta|_c = 1} |\alpha_r \beta_n c_{rn}| \quad [r = 1, \dots, \omega;\ n = 1, \dots, m].$$

Note that for α in R_ω and β in R_m,

$$|\alpha|_p \leqq \omega^{\frac{1}{p}} |\alpha|_c, \qquad |\beta|_q \leqq m^{\frac{1}{q}} |\beta|_c,$$

and that accordingly,

$$(4.23)\qquad |\alpha_r \beta_n c_{rn}| \leq |\alpha|_p \, |\beta|_q \, F_{pq}^{(m)} \leq \omega^{\frac{1}{p}} m^{\frac{1}{q}} \, |\alpha|_c \, |\beta|_c \, F_{pq}^{(m)}.$$

It follows from (4.22) and (4.23) that

$$F^{(m)} \leq \omega^{\frac{1}{p}} m^{\frac{1}{q}} F_{pq}^{(m)}.$$

In particular,

$$F^{(m)} \leq \omega^{\frac{1}{2}} m^{\frac{1}{2}} F_{22}^{(m)} = \omega m^{\frac{1}{2}},$$

using Lemma 4.5. Thus

$$(4.24)\qquad \frac{F^{(m)}}{\omega m} = O(1/m^{\frac{1}{2}}).$$

5. An integral d-function $k^{(m)}$. In terms of the matrix **c** of elements c_{rn} ($r = 1, \ldots, \omega$; $n = 1, \ldots, m$; $\omega = 2^{m-1}$), we shall define a d-function $k^{(m)}$ relative to each product $L_p \times L_q$ ($p > 1$, $q > 1$), and evaluate $h(L_p, L_q; k^{(m)})$ in terms of $F_{pq}^{(m)}$. As previously m is an odd integer.

Let a partition π_m of $E' \times E''$ be defined by a division of E' into ω equal subintervals, and E'' into m equal subintervals. Let R_{rn} be that subrectangle of $E' \times E''$ of dimension $1/\omega$ by $1/m$ which is the Cartesian product of the rth subinterval of E' by the nth subinterval of E''. Let $\theta^{(m)}$ map R_2 into R_1 with $\theta^{(m)}(u,v) = 0$ over the complement of $E' \times E''$ and $\theta^{(m)}(u,v) = c_{rn}$ for (u,v) in R_{rn}. Set

$$(5.1)\qquad k^{(m)}(s,t) = \int_0^s du \int_0^t \theta^{(m)}(u,v)\, dv.$$

One sees that $k^{(m)}$ is left canonical (§2). We shall prove the following lemma.

LEMMA 5.1. The function $k^{(m)}$ is a d-function relative to $L_p \times L_q$ ($p > 1$, $q > 1$) with

$$(5.2)\qquad h(L_p, L_q; k^{(m)}) = \omega^{-\frac{1}{p'}} m^{-\frac{1}{q'}} F_{pq}^{(m)}.$$

In accordance with Lemma 2.1,

$$(5.3) \qquad h(L_p, L_q; k^{(m)}) = \sup_{\eta, \xi} \eta_r \xi_n \Delta_{rn}(k^{(m)}),$$

taking the sup over unit step-functions η and ξ in L_p and L_q respectively, and restricting η and ξ to step-functions associated with the partition π_m of E' X E". It follows from the definition of $k^{(m)}$ that

$$(5.4) \qquad \Delta_{rn}(k^{(m)}) = \frac{c_{rn}}{\omega m} .$$

To obtain the relation (5.2) one considers η and ξ as defining vectors $(\eta_1, \ldots, \eta_\omega)$ and $(\xi_1, \ldots, \xi_m)$ in R_ω and R_m respectively, with norms $|\eta|_p$ and $|\xi|_q$ in R_ω and R_m [Cf. (4.0)]. Recalling that the subintervals of E' and E" under π_m have the lengths $1/\omega$ and $1/m$ respectively, it is seen that

$$\omega^{\frac{1}{p}} |\eta|_{L_p} = |\eta|_p , \qquad m^{\frac{1}{q}} |\xi|_{L_q} = |\xi|_q ,$$

so it follows from (5.3) that for $p > 1$, $q > 1$,

$$h(L_p, L_q; k^{(m)}) = \sup_{\bar\eta, \bar\xi} \bar\eta_r \bar\xi_n \Delta_{rn}(k^{(m)})$$

$$= \sup_{\bar\eta, \bar\xi} \bar\eta_r \bar\xi_n \left(\frac{c_{rn}}{\omega m}\right) \leq \sup_{\bar\eta, \bar\xi} \frac{|\bar\eta|_p \, |\bar\xi|_q F_{pq}^{(m)}}{\omega\, m} = \frac{\omega^{\frac{1}{p}} m^{\frac{1}{q}}}{\omega m} F_{pq}^{(m)} ,$$

taking the sup over vectors $\bar\eta$ in R_ω with $|\bar\eta|_p = \omega^{\frac{1}{p}}$, and vectors $\bar\xi$ in R_m with $|\bar\xi|_q = m^{\frac{1}{q}}$. Relation (5.2) follows.

This lemma taken with Lemma 4.7 gives a basic evaluation.

THEOREM 5.1. For $1 < q \leq 2$ and for $p' = 2, 3, \ldots$,

$$(5.5) \qquad h(L_p, L_q; k^{(m)}) \leq m^{-\frac{2}{p'q'}} \qquad \left[\frac{1}{p} + \frac{1}{p'} = 1, \ \frac{1}{q} + \frac{1}{q'} = 1\right].$$

Relation (5.5) follows from (5.2) and the formula (4.18) for $F_{pq}^{(m)}$. On taking the p'th root of the right member of (4.18) the resulting exponents of ω and m on the right of (5.5) are respectively

$$\frac{1}{p'} - \frac{1}{p'} = 0, \qquad \frac{p' - 2}{p'q'} - \frac{1}{q'} = \frac{-2}{p'q'},$$

so that (5.5) holds.

COROLLARY 5.1. For any $p > 1$ and $q > 1$,

$$\text{(5.6)} \qquad \lim_{m \uparrow \infty} h(L_p, L_q; k^{(m)}) = 0 \qquad \text{[m an odd integer].}$$

In this proof we shall refer to the pairs (p,q) admitted in Theorem 5.1 as <u>special</u>. From (5.5) it follows that (5.6) holds for each special pair (p,q). But given an arbitrary pair (p,q) with $p > 1$, $q > 1$, there is a special pair (μ, ν) such that $p > \mu$ and $q > \nu$. For such a pair $L_p \subset L_\mu$ and $L_q \subset L_\nu$, so that

$$h(L_p, L_q; k^{(m)}) \leq h(L_\mu, L_\nu; k^{(m)})$$

in accordance with Comparison Theorem 3.0. Hence (5.6) holds for each pair (p,q) for which $p > 1$, $q > 1$.

<u>The Vitali variation</u> $V(k^{(m)})$. According to a theorem of Lebesgue, p. 383,

$$\text{(5.7)} \qquad V(k^{(m)}) = \int_{R_1}\int |e^{(m)}(s,t)|\, ds\, dt = \int_0^1 \int_0^1 ds\, dt = 1.$$

COROLLARY 5.2. The function $k^{(m)}$ is a d-function relative to $C \times C$ such that $V(k^{(m)}) = 1$ and

$$\text{(5.8)} \qquad h(C,C;k^{(m)}) \leq m^{-\frac{1}{2}}.$$

By virtue of Comparison Theorem 3.0 and Theorem 5.1,

$$h(C,C;k^{(m)}) \leq h(L_2,L_2;k^{(m)}) \leq m^{-\frac{1}{2}}$$

with $p = q = 2$ in Theorem 5.1. This gives (5.8).

<u>6</u>. d-<u>Functions</u> K_{pq} <u>with</u> $V(K_{pq}) = \infty$. For each $p > 1$ and $q > 1$ we shall define a d-function K_{pq} relative to $L_p \times L_q$ for which $V(K_{pq}) = \infty$,

making use of the elementary d-functions $k^{(m)}$ of §5. For these d-functions the Fréchet variation is finite by definition of a d-function.

Let Q_n $(n = 1, 2, \ldots)$ be a sequence of disjoint squares each in the interval $(0,1] \times (0,1]$ of the (s,t)-plane with diagonals on which $s = t$ and with the maximum distance of the vertices of Q_n from $(0,0)$ tending to 0 as $n \to \infty$. For $n = 1, 2, \ldots$ let Q_n be the image of $[0,1] \times [0,1]$ under the transformation

$$(6.1) \qquad \begin{aligned} u &= a_n s + b_n, \\ v &= a_n t + b_n. \end{aligned} \qquad [0 < b_n < a_n + b_n < 1]$$

Let $m_n > 1$ be an odd integer. Define $g^{(m_n)}$ by setting

$$(6.2) \qquad k^{(m_n)}(s,t) = g^{(m_n)}(u,v), \qquad n = 1, 2, \ldots$$

subject to (6.1). Set

$$(6.3) \qquad K_{pq}(u,v) = g^{(m_1)}(u,v) + g^{(m_2)}(u,v) + \ldots$$

for every point (u,v) in R_2. The integers m_n shall increase with n, and are to be chosen so as to satisfy various conditions depending upon the application. A first irreducible condition is that

$$(6.4) \qquad P(m_1) + P(m_2) + \ldots < \infty,$$

where

$$(6.5) \qquad P(m_n) = h(L_p, L_q; g^{(m_n)}).$$

Observe that $g^{(m_n)}$ is continuous and left canonical relative to $E' \times E''$, and that it follows from Lemma 2.3 that

$$(6.6) \qquad P(m_n) = a_n^{-\frac{1}{p}-\frac{1}{q}} h(L_p, L_q; k^{(m_n)}).$$

Since the coefficient $a_n^{-\frac{1}{p}-\frac{1}{q}}$ is independent of the choice of m_n it follows from Corollary 5.1 that $P(m_n)$ is less than a prescribed constant (such as

$1/n^2$) if m_n exceeds a sufficiently large integer N_m. Hence the series (6.4) converges for suitable choice of the integers m_n. We suppose the integers m_n so chosen.

Since $C \subset L_p$ and $C \subset L_q$ it follows from the Comparison Theorem 3.0 that

$$h(C,C;g^{(m_n)}) \leq h(L_p,L_q;g^{(m_n)}) = P(m_n). \tag{6.7}$$

Since $g^{(m_n)}$ is left canonical relative to $E' \times E''$,

$$g^{(m_n)}(s,t) \leq h(C,C;g^{(m_n)}) \leq P(m_n). \tag{6.8}$$

We conclude that the series on the right of (6.3) converges uniformly, absolutely, and boundedly to a bounded continuous function K_{pq}. Hence K_{pq} is left canonical relative to $E' \times E''$ since its terms $g^{(m_n)}$ are left canonical.

LEMMA 6.1. With the integers m_n chosen so that m_n increases with n and (6.4) is satisfied, the Vitali variation $V(K_{pq}) = \infty$.

With f mapping R_2 into R_1 let $V(Q,f)$ and $P(Q,f)$ respectively denote the Vitali and Fréchet variations of f over the 2-interval Q. It follows from the definition of $k^{(m)}$ [Cf. (5.7)] that when $Q = E' \times E''$, $V(Q, k^{(m)}) = 1$. Since the Vitali variation is invariant under the mapping (6.1),

$$V(Q_n,g^{(m_n)}) = V(Q,g^{(m_n)}) = 1.$$

It is clear that

$$V(Q_n,K_{pq}) = V(Q_n,g^{(m_n)}) = 1,$$

$$V(K_{pq}) \geq \sum_n V(Q_n,K_{pq}) = \sum_n V(Q_n,g^{(m_n)}) = \infty,$$

thus establishing the lemma.

We come to a fundamental theorem.

THEOREM 6.1. Given $p > 1$ and $q > 1$, the function K_{pq} defined by (6.3) with the integers m_n increasing with n and m_n chosen so that (6.4) holds, is such that $V(K_{pq}) = \infty$, and

$$0 < h(L_p,L_q;K_{pq}) < \infty. \tag{6.9}$$

That $V(K_{pq}) = \infty$ has been established in the preceding lemma. It remains to verify (6.9). It follows from Lemma 2.2 and (6.5) respectively, that

$$h(L_p, L_q; K_{pq}) \leq P(m_1) + P(m_2) + \dots .$$

Finally, it is impossible that

$$(6.10) \qquad h(L_p, L_q; K_{pq}) = 0.$$

For (2.7) and (2.8) imply that (6.10) holds only if $K_{pq}(s,t) \equiv 0$, and this is clearly impossible when $V(K_{pq}) = \infty$. (Cf. Theorem 12.1, MT 2).

This completes the proof of the theorem.

The preceding theorem and the Comparison Theorem 3.0 give the following:

THEOREM 6.2. There exists a d-function relative to any product space $A \times B$ of the form

$$(6.11) \qquad C \times C, \qquad C \times L_p, \qquad L_p \times C, \qquad L_p \times L_q, \qquad [p > 1,\ q > 1]$$

for which $V(k) = \infty$, and

$$0 < h(A,B;k) < \infty.$$

The d-function k is non-degenerate in the sense that the functional L-S-integral f defined by (0.1) is not the null functional over $A \times B$.

As established in Theorem 12.1, MT 2, a d-function k relative to $A \times B$ defines a null functional over $A \times B$ of the form (0.1) if and only if $h(A,B;k) = 0$.

To state Theorem 6.3 new notation is required. Let s and t be positive numbers. Let k be a d-function relative to $A \times B$. Let $h_{s,t}(A,B;k)$ be the A-B-variation of k over the rectangle $\begin{bmatrix} s,t \\ 0,0 \end{bmatrix}$ determined by the vertices (0,0) and (s,t), with $h_{s,t}(A,B;k)$ defined as was $h(A,B;k)$, on replacing partitions of E' and E" by partitions of [0,s] and [0,t] respectively. Let $V\begin{bmatrix} a',b' \\ a\ ,b \end{bmatrix}(k)$ be the Vitali variation of k over $\begin{bmatrix} a',b' \\ a,\ b \end{bmatrix}$. We shall examine

$$(6.12) \qquad h_{s,t}(A,B;k)$$

for small s and t and for specially chosen k.

THEOREM 6.3. If $A \times B$ is any one of the products (6.11), and if μ is any positive number, there exists a d-function k relative to $A \times B$ such that

$$h_{s,t}(A,B;k) \leq N(st)^{\mu} \qquad [N \text{ a constant}] \tag{6.13}$$

with $V\begin{bmatrix} s,t \\ 0,0 \end{bmatrix}(k) = \infty$ for each positive s and t.

We refer to the d-function K_{pq} of Theorem 6.1 and to the squares Q_n used in defining K_{pq}. Suppose that $Q_n = \begin{bmatrix} c_n, c_n \\ b_n, b_n \end{bmatrix}$. As previously we suppose that

$$0 < b_{n+1} < c_{n+1} < b_n < c_n < 1 \qquad [n = 1, 2, \ldots].$$

We can and will choose these squares so that

$$\frac{c_n}{c_{n+1}} < M, \qquad [n = 1, 2, \ldots] \tag{6.14}$$

where M is a constant independent of n. Suppose that the odd integers m_n used in defining K_{pq} increase with n and are such that

$$P(m_n) < c_n^{2\mu} - c_{n+1}^{2\mu}. \tag{6.15}$$

This is possible since $P(m) \to 0$ as $m \uparrow \infty$.

By virtue of the Comparison Theorem 3.0, it will be sufficient to prove Theorem 6.3 for the case in which $A = L_p$, $B = L_q$, and to show that K_{pq} can serve as the function k of the theorem. Since p and q are fixed, we write

$$h_{s,t}(L_p,L_q;g) = h_{s,t}(g)$$

for brevity. The square $\begin{bmatrix} c_n, c_n \\ 0\ , 0 \end{bmatrix}$ does not intersect $Q_1, Q_2, \ldots, Q_{n-1}$. It follows that

$$h_{c_n,c_n}(g^{(m_r)}) = 0 \qquad [r = 1, \ldots, n-1].$$

Hence if one sets $g^{(m_n)} + g^{(m_{n+1})} + \ldots = R_n$,

$$(6.16)\qquad h_{c_n,c_n}(K_{pq}) = h_{c_n,c_n}(R_n) \leqq P(m_n) + P(m_{n+1}) + \cdots < c_n^{2\mu},$$

using Lemma 2.2 and (6.15).

Without loss of generality we can suppose that $s \geqq t$. If $b_n \leqq t < b_{n-1}$, the rectangle $\begin{bmatrix} s,t \\ 0,0 \end{bmatrix}$ intersects none of the squares $Q_{n-1}, \ldots, Q_1$, so that

$$h_{s,t}(K_{pq}) \leqq h_{c_n,c_n}(K_{pq}) < c_n^{2\mu} \qquad \text{[by (6.16)]}$$

$$= \left[\frac{c_n}{c_{n+1}}\right]^{2\mu} c_{n+1}^{2\mu} \leqq M^{2\mu} t^{2\mu} \leqq M^{2\mu}(st)^{\mu} \text{ [when } s \geqq t].$$

When $t \geqq s$, a similar result is obtainable. This establishes (6.13).

The proof that $V\begin{bmatrix} s,t \\ 0,0 \end{bmatrix}(K_{pq}) = \infty$ for any positive s and t is similar to the proof of Lemma 6.1. Theorem 6.3 is thereby established.

7. The product $C \times C$. The results obtained for a general product can be simplified and extended in the case of the product $C \times C$. Earlier notation and concepts require extensions.

Let Q be an arbitrary 2-dimensional interval over which a function k is defined. If Q is closed, the Fréchet variation $P(Q,k)$ of k over Q is well defined (Cf. MT 1). If Q is not closed, we introduce the definition

$$(7.1)\qquad P(Q,k) = \sup_J P(J,k),$$

taking the sup over all closed subintervals J of Q. When $Q = \begin{bmatrix} a',b' \\ a\ ,b \end{bmatrix}$, we shall write

$$(7.2)\qquad P(Q,k) = P\begin{bmatrix} a',b' \\ a\ ,b \end{bmatrix}(k).$$

In this section we shall set

$$(7.3)\qquad \Omega = [0,1] \times [0,1], \qquad \Omega_0 = (0,1] \times (0,1].$$

Functions weakly in $L(Q)$. If φ is defined over an interval Q and is in L over every closed subinterval of Q, φ will be said to be weakly in L over Q. If Q is closed, a function weakly in $L(Q)$ is in L over Q. If φ is weakly in $L(\Omega_0)$, we shall set

$$(7.4)\qquad \overline{\Phi}(u,v) = \int_1^u \int_1^v \varphi(s,t)\,ds\,dt \qquad [(u,v) \in \Omega_0].$$

The scope of our counter examples will be considerably increased if the elementary function $k^{(m)}$ of §5 is replaced by a function $f^{(m)}$ defined over Ω as follows. In terms of the function $e^{(m)}$ of (5.1) set

$$(7.5)\qquad K^{(m)}(u,v) = \int_1^u \int_1^v e^{(m)}(s,t)\,ds\,dt.$$

One recognizes at once that

$$(7.6)'\qquad P(\Omega, K^{(m)}) = P(\Omega, k^{(m)}) \leqq m^{-\frac{1}{2}}, \qquad \text{[by Corollary 5.2]}$$

$$(7.6)''\qquad V(\Omega, K^{(m)}) = V(\Omega, k^{(m)}) = 1, \qquad \text{[by (5.7)]}.$$

We now define $f^{(m)}$ over Ω as follows:

$$(7.7)^{I}\qquad f^{(m)}\left[\frac{u+1}{2}, \frac{v+1}{2}\right] = \frac{K^{(m)}(u,v)}{4}, \qquad [0 < u \leqq 1;\ 0 < v \leqq 1]$$

$$(7.7)^{II}\qquad f^{(m)}\left[\frac{u+1}{2}, \frac{v}{2}\right] = \frac{K^{(m)}(u,1-v)}{4}, \qquad [0 < u \leqq 1;\ 0 \leqq v \leqq 1]$$

$$(7.7)^{III}\qquad f^{(m)}\left[\frac{u}{2}, \frac{v+1}{2}\right] = \frac{K^{(m)}(1-u,v)}{4}, \qquad [0 \leqq u \leqq 1;\ 0 < v \leqq 1]$$

$$(7.7)^{IV}\qquad f^{(m)}\left[\frac{u}{2}, \frac{v}{2}\right] = \frac{K^{(m)}(1-u,1-v)}{4}, \qquad [0 \leqq u \leqq 1;\ 0 \leqq v \leqq 1]$$

and set $f^{(m)}(u,v) = 0$ for (u,v) not in Ω. One sees that

$$(7.8)\qquad f^{(m)}(u,v) = \int_1^u \int_1^v \mathfrak{f}^{(m)}(s,t)\,ds\,dt,$$

where $\mathfrak{f}^{(m)}(s,t) = 0$ for (s,t) not in Ω, and

$$(7.9)\qquad \mathfrak{f}^{(m)}\left[\frac{u+1}{2}, \frac{v+1}{2}\right] = e^{(m)}(u,v), \qquad [0 < u \leqq 1;\ 0 < v \leqq 1]$$

with similar relations derived from $(7.7)^{II}$, $(7.7)^{III}$ and $(7.7)^{IV}$. The equalities (7.7) imply the relations

$$(7.10)' \qquad P\begin{bmatrix}1, & 1\\ \frac{1}{2}, & \frac{1}{2}\end{bmatrix}(f^{(m)}) = \frac{1}{4}P\begin{bmatrix}1, & 1\\ 0, & 0\end{bmatrix}(K^{(m)}),$$

$$(7.10)'' \qquad V\begin{bmatrix}1, & 1\\ \frac{1}{2}, & \frac{1}{2}\end{bmatrix}(f^{(m)}) = \frac{1}{4}V\begin{bmatrix}1, & 1\\ 0, & 0\end{bmatrix}(K^{(m)}),$$

with similar relations corresponding to $(7.7)^{II}$, $(7.7)^{III}$, $(7.7)^{IV}$. It follows that

$$(7.11)' \qquad P\begin{bmatrix}1, & 1\\ 0, & 0\end{bmatrix}(f^{(m)}) \leqq P\begin{bmatrix}1, & 1\\ 0, & 0\end{bmatrix}(K^{(m)}) \leqq m^{-\frac{1}{2}},$$

$$(7.11)'' \qquad V\begin{bmatrix}1, & 1\\ 0, & 0\end{bmatrix}(f^{(m)}) = V\begin{bmatrix}1, & 1\\ 0, & 0\end{bmatrix}(K^{(m)}) = 1.$$

We note the following:

<u>The function $f^{(m)}$ vanishes on the boundary and exterior of Ω in</u> R_2.

Let I be an arbitrary closed square, and let

$$(7.12) \qquad \begin{aligned} u &= as + b, \\ v &= at + b, \end{aligned}$$

be a 1-1 mapping of Ω onto I. Under (7.12) set

$$(7.13)' \qquad f^{(m)}(s,t) = \Gamma^{(m)}(I;u,v).$$

As a consequence of (7.13)',

$$(7.13)'' \qquad V(I, \Gamma^{(m)}) = 1, \qquad P(I, \Gamma^{(m)}) \leqq m^{-\frac{1}{2}}.$$

We refer to the sequence of non-intersecting squares,

$$Q_n = \begin{bmatrix}c_n, c_n\\ b_n, b_n\end{bmatrix}, \qquad [n = 1, 2, \ldots]$$

tending to the origin as $n \to \infty$, and state the following theorem summarizing and completing results of §6.

THEOREM 7.1. With $d_n \geq 0$ $(n = 1, 2, \dots)$ and m_n an odd integer increasing with n, suppose

$$\lambda = \frac{d_1}{\sqrt{m_1}} + \frac{d_2}{\sqrt{m_2}} + \dots < \infty. \tag{7.14}$$

Then the series

$$d_1 \Gamma^{(m_1)}(Q_1;u,v) + d_2 \Gamma^{(m_2)}(Q_2;u,v) + \dots = \Gamma(u,v) \tag{7.15)'}$$

defines a continuous function Γ over R_2 such that $\Gamma(u,v) = 0$ for (u,v) not in union Q_n, and

$$\Gamma(u,v) = d_n \Gamma^{(m_n)}(Q_n;u,v), \qquad [(u,v) \in Q_n] \tag{7.15)"}$$

$$V(\Omega, \Gamma) = d_1 + d_2 + \dots = V(\Omega_0, \Gamma), \tag{7.16}$$

$$P(\Omega, \Gamma) \leq \lambda, \qquad \text{[by Lemma 2.2]} \tag{7.17}$$

$$\Gamma(u,v) = \int_1^u \int_1^v \varphi(s,t)\, ds\, dt, \qquad [u \geq 0,\ v \geq 0,\ u+v > 0] \tag{7.18}$$

where φ is weakly in $L(\Omega_0)$, and in $L(\Omega)$ if and only if $d_1 + d_2 + \dots < \infty$.

All statements of the theorem have been amply covered previously with the possible exception of the statement concerning (7.18). However, the definition of $\Gamma^{(m)}$ shows that

$$\Gamma^{(m_n)}(Q_n;u,v) = \int_1^u \int_1^v \varphi_n(s,t)\, ds\, dt,$$

where φ_n may be taken bounded and measurable over Q_n with $\varphi_n(s,t) = 0$ for (s,t) not in Q_n, and where

$$[\text{area } Q_n]\ |\varphi_n(s,t)| = 1 \qquad [(s,t) \in Q_n].$$

One satisfies (7.18) with

$$\varphi(s,t) = d_1 \varphi_1(s,t) + d_2 \varphi_2(s,t) + \dots , \tag{7.19}$$

and observes that for any closed interval I which does not intersect $Q_{n+1} \cup Q_{n+2} \cup \cdots$,

$$(7.20) \quad V(I,\Gamma) = \int_I \int |\varphi(s,t)|\, ds\, dt \leqq d_1 + d_2 + \cdots + d_n,$$

where the equality holds if in addition

$$Q_1 \cup Q_2 \cup \cdots \cup Q_n \subset I.$$

Hence (7.16) holds, and φ is in $L(\Omega)$ if and only if $d_1 + d_2 + \cdots < \infty$.

The following theorem is proved as was Theorem 6.3 replacing $h_{s,t}(L_p,L_q;g)$ by $P\begin{bmatrix} s,t \\ 0,0 \end{bmatrix}(g)$, and $g^{(m_r)}$ by $\Gamma^{(m_r)}$, again using Lemma 2.2.

THEOREM 7.2. There exist indefinite integrals Γ of Theorem 7.1 such that $V\begin{bmatrix} s,t \\ 0,0 \end{bmatrix}(\Gamma) = \infty$ for every $s > 0$, $t > 0$ while

$$(7.21) \quad P\begin{bmatrix} s,t \\ 0,0 \end{bmatrix}(\Gamma) \leqq N(st)^{\mu}$$

for prescribed $N > 0$ and $\mu > 0$.

The Fréchet and Vitali variations are <u>topological invariants</u> with respect to homeomorphisms X and Y of E' and E" in the following sense. Set

$$k(s',t') = g(s,t) \qquad [\text{with } s' = X(s);\ t' = Y(t)].$$

It then follows from the definition of the Vitali and Fréchet variations that

$$(7.22) \quad V\begin{bmatrix} s,t \\ 0,0 \end{bmatrix}(g) = V\begin{bmatrix} s',t' \\ 0,0 \end{bmatrix}(k), \qquad P\begin{bmatrix} s,t \\ 0,0 \end{bmatrix}(g) = P\begin{bmatrix} s',t' \\ 0,0 \end{bmatrix}(k).$$

Making use of this invariance we can now state the following corollary of Theorem 7.2.

COROLLARY 7.1. Let λ and μ be arbitrary continuous functions with $\lambda(s) > 0$, $\mu(s) > 0$ for $s > 0$. There exist indefinite integrals Γ of Theorem 7.1 such that $V\begin{bmatrix} s,t \\ 0,0 \end{bmatrix}(\Gamma) = \infty$ for every $s > 0$, $t > 0$ while

$$(7.23) \quad P\begin{bmatrix} s,t \\ 0,0 \end{bmatrix}(\Gamma) \leqq \lambda(s)\,\mu(t).$$

In proving the next lemma two inequalities will be used. If $Q = \begin{bmatrix} a', b' \\ a, b \end{bmatrix}$ is a closed subinterval of Ω_0, if f_1 and f_2, with values $f_1(s)$ and $f_2(t)$, are continuous over $[a,a']$ and $[b,b']$ respectively, and if $\overline{\varphi}$ is an indefinite integral of the form (7.4), then

$$V(Q, \overline{f_1 f_2 \varphi}) = \int_a^{a'} ds \int_b^{b'} |f_1(s)\, f_2(t)\, \varphi(s,t)|\, ds\, dt$$

$$\leq \|f_1\|\ \|f_2\|\ V(Q, \overline{\varphi}), \tag{7.24}$$

where $\|f_1\|$ is the maximum of $f_1(s)$ over $[a,a']$ and $\|f_2\|$ the maximum of $f_2(t)$ over $[b,b']$. We shall use (7.24) and a similar relation

$$P(Q, \overline{f_1 f_2 \varphi}) \leq \|f_1\|\ \|f_2\|\ P(Q, \overline{\varphi}), \tag{7.25}$$

established in §6, MT 4.

A lemma instrumental in showing that our generalization of Gergen's test is less restrictive than any of the classical 2-dimensional tests is as follows.

LEMMA 7.1. *There exists a function φ in $L(\Omega)$ such that $P(\Omega_0, \overline{\varphi/uv}) < \infty$, and arbitrarily small constants $c > 0$ such that with $\begin{bmatrix} c, c \\ 0+, 0+ \end{bmatrix}$ the interval $(0,c] \times (0,c]$*

$$V \begin{bmatrix} c, c \\ 0+, 0+ \end{bmatrix} (\overline{uv\ \varphi}) \geq c^3, \qquad V \begin{bmatrix} 2c, 2c \\ c, c \end{bmatrix} (\overline{\varphi}) = 0. \tag{7.26}$$

We shall establish this lemma with the aid of Theorem 7.1, using the function φ in Theorem 7.1. To that end set

$$Q_n = \begin{bmatrix} \frac{1}{2^n}, & \frac{1}{2^n} \\ \frac{1}{2^{n+1}}, & \frac{1}{2^{n+1}} \end{bmatrix} = \begin{bmatrix} a_n, & a_n \\ \frac{a_n}{2}, & \frac{a_n}{2} \end{bmatrix}, \qquad [n = 1, 2, \ldots] \tag{7.27}$$

introducing a_n, and setting $d_{2r} = 4a_{2r}$ and $d_{2r-1} = 0$ $(r = 1, 2, \ldots)$ in Theorem 7.1. Choose m_{2r} so large that

$$\frac{d_{2r}}{\sqrt{m_{2r}}} \leq \frac{1}{4} (a_{2r})^3 \qquad [r = 1, 2, \ldots]. \tag{7.27'}$$

We shall show that the resulting function φ of Theorem 7.1 as defined by

(7.19) satisfies the lemma.

From (7.24) for $r = 1, 2, \ldots$,

$$V(Q_{2r}, \overline{uv\,\varphi}) \geqq \left(\frac{a_{2r}}{2}\right)^2 V(Q_{2r}, \overline{\varphi}) = \frac{(a_{2r})^2}{4} d_{2r} = (a_{2r})^3,$$

$$\text{(7.28)} \qquad V \begin{bmatrix} a_{2r}, a_{2r} \\ 0+, 0+ \end{bmatrix} (\overline{uv\,\varphi}) \geqq \sum_{p=r}^{\infty} (a_{2p})^3 = \frac{64}{63} (a_{2r})^3.$$

By virtue of (7.25) and (7.27)' respectively,

$$P(Q_{2r}, \overline{\varphi/uv}) \leqq \left(\frac{2}{a_{2r}}\right)^2 P(Q_{2r}, \overline{\varphi}) \leqq \frac{4}{(a_{2r})^2} \frac{(a_{2r})^3}{4} = a_{2r},$$

$$\text{(7.29)} \qquad P(\Omega_0, \overline{\varphi/uv}) \leqq \sum_{r=1}^{\infty} a_{2r} = \sum_{r=1}^{\infty} \left(\tfrac{1}{4}\right)^r = \tfrac{1}{3}.$$

Since the series $d_1 + d_2 + \ldots$ converges, φ is in L over Ω by Theorem 7.1. According to (7.28) the first relation in (7.26) is satisfied on taking $c = a_{2r}$ $(r = 1, 2, \ldots)$. Moreover,

$$V \begin{bmatrix} 2a_{2r}, 2a_{2r} \\ a_{2r}, a_{2r} \end{bmatrix} (\overline{\varphi}) = V(Q_{2r-1}, \overline{\varphi}) = d_{2r-1} = 0.$$

This establishes the lemma.

In proving the next lemma we shall need a relation established under Lemma 4.2 of MT 4. Let g be defined over the interval $Q = \begin{bmatrix} a', b' \\ a\ , b \end{bmatrix}$, and let k, with values k(t), be defined over [b,b'] with total Jordan variation $T_b^{b'}(k)$. Then with $g(s,b) \equiv g(a,t) \equiv 0$

$$\text{(7.30)} \qquad P(Q,kg) \leqq [\sup_t |k(t)| + T_b^{b'}(k)]\, P(Q,g) \qquad [b \leqq t \leqq b'].$$

In particular if $0 < b < b'$,

$$P(Q, \tfrac{g}{t}) \leqq \left[\frac{1}{b} + \left(\frac{1}{b} - \frac{1}{b'}\right)\right] P(Q,g) \leqq \frac{2}{b} P(Q,g).$$

Applying (7.30) with the roles of s and t interchanged, and supposing that $0 < b < b'$, $0 < a < a'$, one infers that

(7.31) $$P(Q, \frac{g}{st}) \leq \frac{4}{ab} P(Q, g).$$

The following lemma will permit us to show that there are functions φ which satisfy our 2-dimensional Jordan test but do not satisfy the Young-Pollard test (Y) as defined by Gergen.

LEMMA 7.2. There exists a function g continuous over Ω, vanishing on the boundary of Ω with $P(\Omega, g) < \infty$, and such that for every $x \in (0,1]$, $V\begin{bmatrix} x,x \\ 0,0 \end{bmatrix}(uvg) = \infty$.

For $u > 0$, $v > 0$, $g(u,v)$ will be taken of the form $g(u,v) = \Gamma(u,v)/uv$, where Γ is given by Theorem 7.1 with $d_n = 1$, $b_n = 1/n$, for $n = 1, 2, \ldots$, and the integers m_n so chosen that

(7.32) $$\sum_n \frac{1}{b_n^2 \sqrt{m_n}} = \sum_n \frac{n^2}{\sqrt{m_n}} < \infty \qquad [n = 1, 2, \ldots].$$

To satisfy (7.32) one can in particular set $m_n = n^8$.

It follows from Theorem 7.1 that

$$\infty = V\begin{bmatrix} x,x \\ 0,0 \end{bmatrix}(\Gamma) \qquad [x \in (0,1]].$$

Moreover,

$$P(\Omega_0, g) \leq \sum_n P(Q_n, \Gamma/uv) \qquad \text{[from (7.15)']}$$

$$= \sum_n P(Q_n, \Gamma^{(m_n)}/uv) \leq \sum_n 4n^2 P(Q_n, \Gamma^{(m_n)}) \qquad \text{[by (7.31)]}$$

$$\leq 4 \sum_n \frac{n^2}{\sqrt{m_n}} < \infty \qquad \text{[from (7.32)]}.$$

Since $P(\Omega_0, g) < \infty$, it follows from Theorem 5.1, MT 1, that the quadrant limit $g(0+,0+)$ exists. Noting that $\Gamma(u,v)$, and hence $g(u,v)$, vanishes for every point $u > 0$, $v > 0$, not on some square Q_n, we conclude that g admits a continuous extension over Ω such that g vanishes on the boundary of Ω. It follows from Theorem 6.4, MT 5 (cited in the preceding paper) that

$$P(\Omega_0, g) = P(\Omega, g).$$

This establishes the lemma.

LEMMA 7.3. There exists a g in $L(\Omega)$ such that g/uv is not in $L(\Omega)$, while

$$(7.33) \qquad P(\Omega_0, \overline{g/uv}) < \infty.$$

In Theorem 7.1, let $d_n = 1$ and $c_n = 1/n$ $(n = 1, 2, \ldots)$. In terms of the resulting function φ of (7.18), set $g(u,v) = uv\,\varphi(u,v)$. Then

$$(7.34) \qquad P(\Omega_0, \overline{g/uv}) = P(\Omega_0, \overline{\varphi}) \leqq P(\Omega, \Gamma) < \infty,$$

by (7.17), while by (7.16),

$$V(\Omega_0, \overline{g/uv}) = V(\Omega_0, \overline{\varphi}) = \infty.$$

Moreover,

$$V(Q_n, \bar{g}) = V(Q_n, \overline{uv\,\varphi}) \leqq c_n^2\, V(Q_n, \overline{\varphi}) = c_n^2,$$

using (7.24) to remove the factors uv. Hence

$$V(\Omega_0, \bar{g}) = c_1^2 + c_2^2 + \ldots < \infty,$$

thereby establishing the lemma.

According to the lemma,

$$(7.35) \qquad \int_{\Omega_0}\int \left|\frac{g(u,v)}{uv}\right| du\, dv = \infty$$

while $P(\Omega_0, \overline{g/uv}) < \infty$. This situation is typical of comparisons between Vitali and Fréchet variations. It has no strict 1-dimensional counterpart.

References

S. Banach

1. "Théorie des opérations linéaires," Warsaw, 1932.

J. A. Clarkson and C. R. Adams

1. On definitions of bounded variation for functions of two variables, "Trans. Amer. Math. Soc.," vol. 35 (1933), pp. 824-854.

References (cont.)

J. J. Gergen

1. Convergence criteria for double Fourier series, "Trans. Amer. Math. Soc.," vol. 35 (1933), pp. 29-63.

G. H. Hardy

1. "Inequalities," Cambridge Univ. Press, 1934.

T. H. Hildebrandt

1. On integrals related to and extensions of the Lebesgue integrals, "Bull. Amer. Math. Soc.," vol. 24 (1917-18), pp. 177-202.

H. Lebesgue

1. Sur l'integration de fonctions discontinues, "Ann. École Norm. 3," vol. 27 (1910).

P. Lévy

1. Sur les fonctionnelles bilinéaires, "C. R. Acad. Sci. Paris," vol. 222 (Jan. 7, 1946), pp. 125-127.

J. E. Littlewood

1. On bounded bilinear forms in an infinite number of variables, "Quart. J. Math.," Oxford Ser. vol. 1 (1930), pp. 164-174.

E. J. McShane

1. "Integration;" Princeton University Press, 1944.

M. Morse and W. Transue

1. Functionals of bounded Fréchet variation, "Canadian Jour. Math.," vol. 1 (1949), pp. 153-165.

2. Functionals F bilinear over the product $A \times B$ of two pseudo-normed vector spaces, I. The representation of F, "Ann. of Math.," vol. 50 (1949).

3. Functionals F bilinear over the product $A \times B$ of two pseudo-normed vector spaces, II. Admissible spaces A, "Ann. of Math.," vol. 51 (1950).

4. A calculus for Fréchet variations, "Jour. Indian Math. Soc.," (1950).

F. Riesz

1. Untersuchungen über Systeme integrierbarer Funktionen, "Math. Ann.," vol. 69 (1910), pp. 449-497.

A. Zygmund

1. "Trigonometrical series," Warsaw, 1935.

V. NOTE ON THE BOUNDARY VALUES OF FUNCTIONS OF SEVERAL COMPLEX VARIABLES

By A. P. Calderón[1] and A. Zygmund

1. Statement of results. In what follows we shall mostly consider functions $f(z_1, \ldots, z_k)$ of k complex variables $z_1, \ldots, z_k$, defined and regular in the unit polycylinder

$$(\Gamma_k) \qquad |z_1| < 1, \ldots, |z_k| < 1.$$

We shall discuss the problem of the existence of boundary values of such functions f. Instead, however, of the whole boundary of the polycylinder Γ_k, we shall consider only part of it, namely the so called "distinguished boundary," defined by the equations

$$(D_k) \qquad z_1 = e^{i\theta_1}, \ldots, z_k = e^{i\theta_k} \qquad (0 \leqq \theta_1, \ldots, \theta_k \leqq 2\pi)$$

The distinguished boundary D_k is a k-dimensional variety.

We shall say that the function $f(z_1, \ldots, z_k)$ has a non-tangential limit s at a point $(e^{i\theta_1}, \ldots, e^{i\theta_k})$ of D_k, if f tends to s as each of the variables z_j tends, non-tangentially, from within the unit circle $|z_j| < 1$, to $e^{i\theta_j}$. By limit, we shall always mean a finite limit only. It is well known that, if $f(z_1, \ldots, z_k)$ is bounded in Γ_k, then f has a non-tangential limit at almost every point $(e^{i\theta_1}, \ldots, e^{i\theta_k})$. (See [8], where the result is stated explicitly for the radial approach only; the proof, however, is perfectly general). In other words, the set of the points $(\theta_1, \ldots, \theta_k)$ situated in the k dimensional interval

$$(Q_k) \qquad 0 \leqq \theta_1 \leqq 2\pi, \ldots, 0 \leqq \theta_k \leqq 2\pi,$$

and for which the non-tangential limit of f does not exist, is of Lebesgue k-dimensional measure zero.

The same conclusion holds, if the boundedness of f in Γ_k is replaced by the condition that f belongs to a Hardy class, that is if, for some positive α, the integral

1. Fellow of the Rockefeller Foundation.

$$(1) \qquad \int_0^{2\pi}\cdots\int_0^{2\pi} |f(r_1e^{i\theta_1}, \ldots, r_ke^{i\theta_k})|^{\alpha}\, d\theta_1\ldots d\theta_k$$

remains bounded for $r_1 < 1, \ldots, r_k < 1$. Still more generally, the assumption of the boundedness of the integral (1) can be replaced by the condition

$$(2) \qquad \int_0^{2\pi}\cdots\int_0^{2\pi} \log^+ |f| \,(\log^+ \log^+ |f|)^{k-1}\, d\theta_1\ldots d\theta_k = O(1).$$

Here f stands for $f(r_1 e^{i\theta_1}, \ldots, r_k e^{i\theta_k})$ and, as usual, $\log^+ u = \mathrm{Max}\,(\log u, 0)$ for u non-negative. For the proof, see [10].

For $k = 1$ the iterated logarithm in (2) drops out and the result reduces to the very familiar Nevanlinna - Ostrowski theorem. The meaning of the condition (2) becomes easier to understand, if one takes into account another well known fact, this time from the theory of integration, which asserts that if $g(x_1, \ldots, x_k)$ is a measurable function of k real variables $x_1, \ldots, x_k$, such that $|g|(\log^+ |g|)^{k-1}$ is Lebesgue integrable, then the integral of g is <u>strongly differentiable</u> at almost every point $(x_1^0, \ldots, x_k^0)$, (See [4]). In other words, for almost every point $(x_1^0, \ldots, x_k^0)$, if Q denotes a k-dimensional interval with sides parallel to the axes and comprising that point, the limit of

$$\frac{1}{|Q|} \int_Q g\, dx_1\ldots dx_k$$

exists as the diameter of Q tends to 0. Here and hereafter, by $|E|$ we denote the measure of any measurable set E. If in (2) we formally replace $\log^+ |f|$ by g, the condition (2) can be written

$$\int_0^{2\pi}\cdots\int_0^{2\pi} g.(\log^+ g)^{k-1}\, d\theta_1\ldots d\theta_k = O(1),$$

and so takes the form guaranteeing strong differentiability of integrals. As a matter of fact, the latter property and the theorem asserting the existence of the non-tangential limit of the function $f(z_1, \ldots, z_k)$ satisfying condition (2), have a common basic source, namely the Hardy - Littlewood Maximum Theorem (see [3], or [7], p. 241 sqq.).

It is well known that for the strong differentiability of the integral of $g(x_1, \ldots, x_k)$ the condition of the integrability of $|g|(\log^+ |g|)^{k-1}$ is the best possible and cannot be relaxed, (see [6], [4]). It is an open problem whether or not for the existence of the non-tangential limit of $f(z_1, \ldots, z_k)$ the condition (2) is the best possible, but it appears likely that also this condition cannot be relaxed. In particular, one would not expect that the condition

$$(3) \qquad \int_0^{2\pi}\cdots\int_0^{2\pi} \log^+ |f(r_1 e^{i\theta_1}, \ldots, r_k e^{i\theta_k})|\, d\theta_1 \ldots d\theta_k = O(1),$$

which is a straight extension of the Nevanlinna - Ostrowski condition to the case $k > 1$, guarantees the existence of the non-tangential limit of f at almost every point $(e^{i\theta_1}, \ldots, e^{i\theta_k})$.

Still the condition (3) is known to imply the existence almost everywhere of some sort of boundary values. For it can be shown (see [10]) that, if (3) holds then, for almost every θ_1, the non-tangential limit

$$(4) \qquad f^{\theta_1}(z_2, \ldots, z_k) = \lim_{z_1 \to e^{i\theta_1}} f(z_1, z_2, \ldots, z_k)$$

exists uniformly in every polycylinder.

$$|z_2| \leqq 1 - \delta, \ \ldots, \ |z_k| \leqq 1 - \delta \qquad (0 < \delta < 1).$$

Hence $f^{\theta_1}(z_2, \ldots, z_k)$ is, for almost every θ_1, a regular function of $z_2, \ldots, z_k$ in the polycylinder $|z_2| < 1, \ldots, |z_k| < 1$. But more than that is true, namely, for almost every θ_1, the integral

$$\int_0^{2\pi}\cdots\int_0^{2\pi} \log^+ |f^{\theta_1}(r_2 e^{i\theta_2}, \ldots, r_k e^{i\theta_k})|\, d\theta_2 \ldots d\theta_k$$

is bounded as a function of $r_2, \ldots, r_k$ ($r_2 < 1, \ldots, r_k < 1$). Hence, repeating the previous argument, we see that, for almost every θ_1, the non-tangential limit

$$f^{\theta_1\theta_2}(z_3, \ldots, z_k) = \lim_{z_2 \to e^{i\theta_2}} f^{\theta_1}(z_2, \ldots, z_k)$$

(a) exists; (b) is a regular function of $z_3, \ldots, z_k$ in the polycylinder $|z_3| < 1, \ldots, |z_k| < 1$; (c) satisfies the condition

$$\int_0^{2\pi}\cdots\int_0^{2\pi} \log^+ |f^{\theta_1\theta_2}(r_3 e^{i\theta_3}, \ldots, r_k e^{i\theta_k})|\, d\theta_3 \ldots d\theta_k = O(1)$$

-all for almost every θ_2. Thus conditions (a), (b), (c) are satisfied for almost every point (θ_1, θ_2) (almost every, in the two-dimensional sense). Proceeding in this way we see that, under the condition (3), the <u>iterated</u> limit

$$f^{\theta_1 \theta_2 \ldots \theta_k} = \lim_{z_k \to e^{i\theta_k}} f^{\theta_1 \ldots \theta_{k-1}}(z_k)$$

exists at almost every point $(\theta_1, \ldots, \theta_k)$. It can easily be shown (see [10]) that, if the condition (2) is satisfied, then at almost every point $(\theta_1 \ldots \theta_k)$ the limit $f^{\theta_1 \ldots \theta_k}$ is independent of the order in which the variables $z_1, \ldots, z_k$ tend to their respective limits (in particular, this is valid if f belongs to any Hardy class). Whether the assertion remains valid for f satisfying condition (3), seems to be an open problem.

The main purpose of this note is to show that, under condition (3), the function $f(z_1, \ldots, z_k)$ has, at almost every point $(e^{i\theta_1}, \ldots, e^{i\theta_k})$ a so called <u>restricted non-tangential</u> limit. By this we mean that $f(z_1, \ldots, z_k)$ has a limit as each $z_j = r_j e^{i\theta_j}$ tends non-tangentially to $e^{i\theta_j}$, <u>in such a way that all the ratios</u>

$$(4) \qquad \frac{1 - r_j}{1 - r_k} \qquad (j, h = 1, \ldots, k)$$

<u>remain bounded</u>.

The notion of restricted convergence has proved useful and important in the theory of multiple Fourier series. For it is well known that if $g(\theta_1, \ldots, \theta_k)$ is of period 2π in each variable θ_j, and is Lebesgue integrable, and if

$$(5) \qquad g(\theta_1, \ldots, \theta_k) \sim \sum_{n_1, \ldots, n_k = -\infty}^{+\infty} c_{n_1 \ldots n_k} e^{i(n_1\theta_1 + \ldots + n_k\theta_k)}$$

then the Abel means

$$(6) \qquad \sum c_{n_1 \ldots n_k} e^{i(n_1\theta_1 + \ldots + n_k\theta_k)} r_1^{|n_1|} \ldots r_k^{|n_k|}$$

$(r_1 < 1, \ldots, r_k < 1)$ of the Fourier series of g have a restricted non-tangential limit at almost every point $(e^{i\theta_1}, \ldots, e^{i\theta_k})$. If we omit the condition of the boundedness of the ratios (4), the Abel means (5) may diverge everywhere. See [4], [5], and the literature there given.

Of course, restricted Abel summability of a Fourier series reminds one of the classical result of Lebesgue asserting that the integral of any L-integrable function $g(x_1, \ldots, x_k)$ is almost everywhere restrictedly (some say, regularly) differentiable, that is that at almost every point $(x_1^0, \ldots, x_k^0)$ the limit

$$\frac{1}{|Q|} \int_Q g(x_1, \ldots, x_k)\, d\, x_1 \ldots d\, x_k$$

exists as the diameter of Q tends to zero, where Q is any k-dimensional interval containing the point and such that the ratio of any two dimensions of Q remains below a fixed limit.

Thus our main theorem is as follows.

THEOREM 1. If $f(z_1, \ldots, z_k)$ is regular in Γ_k and satisfies condition (3), then at almost every point $(e^{i\theta_1}, \ldots, e^{i\theta_k})$ the function f has a restricted non-tangential limit.

It will be a comparatively simple matter to show that under condition (3) the function f is almost everywhere restrictedly non-tangentially bounded. In other words, for almost every point $(e^{i\theta_1^0}, \ldots, e^{i\theta_k^0})$ the function f remains bounded as $z_j = r_j e^{i\theta_j}$ tends non-tangentially to $e^{i\theta_j^0}$, provided the ratios (4) remain bounded. The proof that f has actually a limit almost everywhere is already deeper.

THEOREM 2. Let $F(z_1, \ldots, z_k)$ be a regular function defined in the polycylinder Γ_k. Suppose that there is a set E situated on the distinguished boundary D_k of Γ_k and having the following property. For every point $(e^{i\theta_1^0}, \ldots, e^{i\theta_k^0})$ in E, the function F remains bounded as $(z_1, \ldots, z_k)$ approaches $(e^{i\theta_1^0}, \ldots, e^{i\theta_k^0})$ restrictedly and non-tangentially. Then F has a restricted non-tangential limit at almost every point of E.

Theorem 2 is of independent interest, but we shall primarily use it in the proof of Theorem 1. It is obvious that the latter theorem is an immediate corollary of Theorem 2 and of the just quoted restricted non-tangential boundedness of functions f satisfying the assumptions of Theorem 1.

In the next section we give proofs of Theorems 1 and 2. The remaining section of the paper is devoted to some additional remarks.

2. Proofs of Theorems 1 and 2. We begin by discussing a class of functions wider than that of regular functions. Let us consider any function $F(z_1, \ldots, z_k)$ defined in the polycylinder Γ_k. We shall say that F is

<u>multiply harmonic in</u> Γ_k, more precisely - k - <u>harmonic there</u>, if it is continuous in Γ_k, with respect to all variables simultaneously, and if it is harmonic in each of the variables z_j separately. In other words, if

$$z_j = x_j + i\, y_j,$$

then $F_{x_j x_j} + F_{y_j y_j} = 0$ for all j. The function F may be complex-valued. If $f(z_1, \ldots, z_k)$ is regular in Γ_k, then f as well as its real and imaginary parts, are k-harmonic. If F is k-harmonic, then it is also a harmonic function of the 2k variables $x_1, y_1, \ldots, x_k, y_k$, the converse being obviously false. We shall also consider k-harmonic functions in domains other than Γ_k, in particular we shall consider them in the topological products V_k of the upper half-planes $y_j > 0$ $(j = 1, \ldots, k)$. Thus V_k is the set of the points $(z_1, \ldots, z_k)$ such that the imaginary parts of all the z_j are positive. If $z_j = h_j(w_j)$ is a linear (fractional) mapping of the unit circle $|z_j| < 1$ onto the upper half-plane $I\, w_j > 0$, and if $F(z_1, \ldots, z_k)$ is k-harmonic in Γ_k, then $F(h_1(w_1), \ldots, h_k(w_k))$ is k-harmonic in V_k, and conversely.

Let us consider any trigonometric series

$$(1) \qquad \sum_{n_1, \ldots, n_k = -\infty}^{+\infty} c_{n_1 \ldots n_k}\, e^{i(n_1\theta_1 + \ldots + n_k\theta_k)}$$

in k real variables $\theta_1, \ldots, \theta_k$. Let us write $z_j = r_j\, e^{i\theta_j}$, and let us assume that the Abel means

$$(2) \qquad F(z_1, \ldots, z_k) = \sum c_{n_1 \ldots n_k}\, e^{i(n_1\theta_1 + \ldots + n_k\theta_k)}\, r_1^{|n_1|} \cdots r_k^{|n_k|}$$

of the series (1) <u>exist</u> for $r_1 < 1, \ldots, r_k < 1$. By this we mean that the series (2) converges absolutely (and so uniformly in $\theta_1, \ldots, \theta_k$) for every system of such values of $r_1, \ldots, r_k$. Then the function $F(z_1, \ldots, z_k)$ is k-harmonic in Γ_k. Conversely, any function $F(z_1, \ldots, z_k)$ which is k-harmonic in Γ_k admits of a representation (2), where the series on the right is absolutely convergent for $r_1 < 1, \ldots, r_k < 1$. For let us fix positive numbers $r_1, \ldots, r_k$ less than 1 and let us consider the periodic function $F(r_1\, e^{i\theta_1}, \ldots, r_k\, e^{i\theta_k})$. The Fourier series of this function is

$$(3) \qquad \sum c_{n_1 \ldots n_k}\, (r_1, \ldots, r_k)\, e^{i(n_1\theta_1 + \ldots + n_k\theta_k)},$$

and the Fourier coefficients are given by the formulas

$$(4)\quad C_{n_1\dots n_k}(r_1,\dots,r_k) = (2\pi)^{-k}\int_0^{2\pi}\dots\int_0^{2\pi} F(r_1 e^{i\theta_1}, \dots, r_k e^{i\theta_k})\, e^{-i(n_1\theta_1+\dots+n_k\theta_k)}\, d\theta_1\dots d\theta_k$$

Since for every harmonic function $H(z)$ of a single variable z, $|z| < 1$, and for every integer m, the integral

$$\int_0^{2\pi} H(r e^{i\theta})\, r^{-|m|}\, e^{-im\theta}\, d\theta$$

is independent of r, $r < 1$, a simple induction (with respect to k) applied to (4) shows that $C_{n_1\dots n_k}(r_1, \dots, r_k)\, r^{-|n_1|}\dots r^{-|n_k|}$ is independent of $r_1, \dots, r_k$. Hence

$$(5)\qquad C_{n_1\dots n_k}(r_1, \dots, r_k) = c_{n_1\dots n_k}\, r_1^{|n_1|}\dots r^{|n_k|},$$

where $c_{n_1\dots n_k}$ are constants. From (5) and (4) one sees that, if we fix positive numbers $R_1, \dots, R_k$ less than 1, then

$$|c_{n_1\dots n_k}| \leqq \text{Const.}\ \frac{1}{R_1^{n_1}\dots R_k^{n_k}}$$

This implies that the series (3), which can now be written

$$\sum c_{n_1\dots n_k}\, e^{i(n_1\theta_1+\dots+n_k\theta_k)}\, r_1^{|n_1|}\dots r_k^{|n_k|}$$

converges absolutely in Γ_k and represents the function $F(z_1, \dots, z_k)$ there.

If (1) is a Fourier series, the function F of (2) is k-harmonic in Γ_k. The same holds, if the series (1) is a Fourier-Stieltjes series, that is if the coefficients $c_{n_1\dots n_k}$ are given by the formulas

$$(6)\qquad c_{n_1\dots n_k} = \frac{1}{(2\pi)^k}\int_0^{2\pi}\dots\int_0^{2\pi} e^{-i(n_1u_1+\dots+n_ku_k)}\, d\mu(E),$$

where $\mu(E)$ is a totally additive, not necessarily positive (or even real-valued) mass distribution defined for all the Borel sets E of the interval

$$(Q_k) \qquad 0 \leqq u_1 \leqq 2\pi, \ \ldots, \ 0 \leqq u_k \leqq 2\pi.$$

Since we identify the opposite faces of the interval Q_k, it would be the same thing to assume that the masses $\mu(E)$ are all confined to the interval

$$0 \leqq u_1 < 2\pi, \ \ldots, \ 0 \leqq u_k < 2\pi.$$

Let $P(r,u)$ denote the Poisson kernel, that is

$$P(r, u) = \frac{1}{2} \sum_{n=-\infty}^{+\infty} r^{|n|} e^{inu} = \frac{1}{2} \frac{1 - r^2}{1 - 2r \cos u + r^2}$$

If we substitute (6) into (2), we get

$$(7) \quad F(r_1 e^{i\theta_1}, \ldots, r_k e^{i\theta_k}) = \pi^{-k} \int_{Q_k} P(r_1, \theta_1 - u_1) \ldots P(r_k, \theta_k - u_k) \, d\mu(E)$$

and the right side here is usually called the (multiple) Poisson-Stieltjes integral of the mass distribution $\mu(E)$.

LEMMA 1. A necessary and sufficient condition that a function $F(z_1, \ldots, z_k)$, which is k-harmonic in Γ_k, be a Poisson- Stieltjes integral is that

$$(8) \qquad \int_{Q_k} |F(r_1 e^{i\theta_1}, \ldots, r_k e^{i\theta_k})| \, d\theta_1 \ldots d\theta_k \leqq M < \infty,$$

with M independent of the values of $r_1, \ldots, r_k$.

The result is very familiar for $k = 1$ (see e.g. [7], p. 86), and the proof for $k > 1$ follows essentially the same line. Since, however, in the latter case we cannot give any reference, a brief sketch of the proof may be desirable. The necessity of the condition is immediate since (7) implies

$$|F(r_1 e^{i\theta_1}, \ldots, r_k e^{i\theta_k})| \leqq \pi^{-k} \int_{Q_k} P(r_1, \theta_1 - u_1) \ldots P(r_k, \theta_k - u_k) | \, d\mu(E)|$$

and so if we integrate this inequality with respect to $\theta_1, \ldots, \theta_k$ over Q_k, interchanging the order of integration on the right, and use the fact that $\pi^{-1}\int_0^{2\pi} P(r,\theta)\,d\theta = 1$, we get (8) with $M = \int_{Q_k} |d\mu|$. Conversely, let us suppose that a k-harmonic function $F(z_1, \ldots, z_k)$ satisfies in Γ_k condition (8). Let $\mu_{r_1 \ldots r_k}(E)$ be the mass distribution over Q_k with density $F(r_1 e^{i\theta_1}, \ldots, r_k e^{i\theta_k})$. Thus

$$(9) \qquad \mu_{r_1 \ldots r_k}(E) = \int_E F(r_1 e^{i\theta_1}, \ldots, r_k e^{i\theta_k})\, d\theta_1 \ldots d\theta_k$$

and, by (8),

$$(10) \qquad \int_{Q_k} |d\mu_{r_1 \ldots r_k}(E)| = \int_{Q_k} |F(r_1 e^{i\theta_1}, \ldots, r_k e^{i\theta_k})|\, d\theta_1 \ldots d\theta_k \leq M .$$

We know that F is representable in the form (2), with

$$(11) \qquad \begin{aligned} c_{n_1 \ldots n_k} r^{|n_1|} \ldots r_k^{|n_k|} &= (2\pi)^{-k} \int_{Q_k} F(r_1 e^{i\theta_1}, \ldots)\, e^{-i(n_1\theta_1 + \ldots)}\, d\theta_1 \ldots d\theta_k \\ &= (2\pi)^{-k} \int_{Q_k} e^{-i(n_1\theta_1 + \ldots + n_k\theta_k)}\, d\mu_{r_1 \ldots r_k}(E) \end{aligned}$$

The masses $\mu_{r_1 \ldots r_k}(E)$ are uniformly bounded, and so we can select from them a subsequence converging to a mass distribution. More precisely, we can find k sequences $\{r_{1,m}\}, \{r_{2,m}\}, \ldots, \{r_{k,m}\}$ $(m = 1, 2, 3, \ldots)$ each tending to 1, and a mass distribution $\mu(E)$ such that for every k-dimensional interval $I \subset Q_k$ we have

$$\mu(I) = \lim_{m \to \infty} \mu_{n_{1,m} \ldots n_{k,m}}(I)$$

provided the absolute mass of μ on the boundary of I is 0 (see [2]). Since the exponential function is continuous, substituting for $r_1, \ldots, r_k$ in (11) the values $r_{1,m}, \ldots, r_{k,m}$, and making m tend to $+\infty$, we obtain from (11) the formula (6). Hence F satisfies the relation (7).

The case of a non-negative mass distribution μ is of importance.

LEMMA 2. A necessary and sufficient condition for a k-harmonic function $F(z_1, \ldots, z_k)$ in Γ_k to be the Poisson - Stieltjes integral of a non-negative mass distribution is that

$$F(z_1, \ldots, z_k) \geq 0 \quad \text{in } \Gamma_k.$$

If μ is non-negative, the formula (7) shows that $F(z_1, \ldots, z_k) \geq 0$. Conversely, if the latter condition is satisfied, then the termwise integration of (2) over Q_k shows that

$$\int_{Q_k} |F|\, d\theta_1 \ldots d\theta_k = \int_{Q_k} F\, d\theta_1 \ldots d\theta_k = (2\pi)^k c_{0\ldots 0} .$$

Hence F is given by the formula (7), and as the proof of Lemma 1 shows, the function $\mu_{r_1 \ldots r_k}(E)$, and so also $\mu(E)$, is non-negative.

LEMMA 3. If $F(z_1, \ldots, z_k)$ is a Poisson - Stieltjes integral in Γ_k, then F has a restricted non-tangential limit at almost every point $(e^{i\theta_1}, \ldots, e^{i\theta_k})$.

For the proof, see [9]. The limit in question is, almost everywhere, the derivative (density) of the mass distribution μ, but this is not necessary for our purposes.

LEMMA 4. Let $f(z_1, \ldots, z_k)$ be a regular function in the unit polycylinder

$$(\Gamma_k) \qquad |z_1| < 1, \ldots, |z_k| < 1 ,$$

and suppose that

$$(12) \qquad \int_0^{2\pi} \cdots \int_0^{2\pi} \log^+ |f(r_1 e^{i\theta_1}, \ldots, r_k e^{i\theta_k})|\, d\theta_1 \ldots d\theta_k = O(1)$$

for $r_1 < 1, \ldots, r_k < 1$. Then there is in Γ_k a k-harmonic function $F(z_1, \ldots, z_k)$ majorizing $\log^+ |f|$ there. Hence, in particular, F is non-negative.

For let us consider the Poisson integral

$$(13) \quad F_{R_1 \ldots R_k}(z_1, \ldots, z_k) = \pi^{-k} \int_{Q_k} \log^+ |f(R_1 e^{iu_1}, \ldots, R_k e^{iu_k})| \prod_{j=1}^{k} P\left(\frac{r_j}{R_j}, u_j - \theta_j\right) du_1 \ldots du_k .$$

The right side here is k-harmonic in the polycylinder $|z_1| < R_1, \ldots |z_k| < R_k$. Moreover, it is continuous in the closure of the polycylinder and coincides with $\log^+|f|$ on its boundary. The function $\log^+|f|$ is a subharmonic function of each variable z_j. Hence the integral

$$(14) \qquad \pi^{-1} \int_0^{2\pi} \log^+|f(r_1 e^{iu_1}, \ldots) P(\frac{r_1}{R_1}, u_1 - \theta_1)\, du_1$$

is a non-decreasing function of R_1, provided $r_1 < R_1 < 1$, for every fixed set of values represented by dots. (This integral is the least harmonic majorant of the subharmonic function $\log^+|f(z_1, \ldots)|$ in the circle $|z_1| \leq R_1$, and it is a very familiar - and elementary - fact that this majorant increases with R_1). If we multiply (14) by

$$P(\frac{r_2}{R_2}, u_2 - \theta_2) \ldots P(\frac{r_k}{R_k}, u_k - \theta_k)$$

and integrate over $(0, 2\pi)$ with respect to all variables $u_2, \ldots, u_k$, we still get a non-decreasing function of R_1. It follows that $F_{R_1 \ldots R_k}(z_1, \ldots, z_k)$ is for a fixed point $(z_1, \ldots, z_k)$ in Γ_k a non-decreasing function of each of the variables $R_1, \ldots, R_k$, provided $|z_1| < R_1, \ldots, |z_k| < R_k$. Hence at every point in Γ_k, the functions $F_{R_1 \ldots R_k}(z_1, \ldots, z_k)$, which are defined for subscripts sufficiently close to 1, tend to a definite limit $F(z_1, \ldots, z_k)$ as $R_1 \to 1, \ldots, R_k \to 1$. This limit is finite since, if all the z_j are numerically less than R, and if all the R_j exceed a number R' such that $R < R' < 1$, then (13) shows that

$$|F_{R_1 \ldots R_k}(z_1, \ldots, z_k)| \leq \left(\frac{1}{2\pi} \frac{R' + R}{R' - R}\right)^k \int_{Q_k} \log^+ |f|\, du_1 \ldots du_k,$$

and so the left side is uniformly bounded. A similar argument, under the same assumptions concerning the z_j and the R_j, shows that the partial derivatives of $F_{R_1 \ldots R_k}(z_1, \ldots, z_k)$ with respect to the r_j and the θ_j are also uniformly bounded, and so the functions $F_{R_1 \ldots R_k}(z_1, \ldots, z_k)$ are uniformly continuous. Hence the convergence of $F_{R_1 \ldots R_k}(z_1, \ldots, z_k)$ to $F(z_1, \ldots, z_k)$ is uniform in every polycylinder $|z_1| \leq R < 1, \ldots, |z_k| \leq R < 1$, which shows that $F(z_1, \ldots, z_k)$ is k-harmonic in Γ_k.

Finally, since $\log^+ |f(z_1, \ldots)|$ is subharmonic in z_1,

$$\log^+ |f(r_1 e^{i\theta_1}, z_2, \ldots, z_k)|$$

$$\leq \pi^{-1} \int_0^{2\pi} \log^+ |f(R_1 e^{iu_1}, z_2, \ldots, z_k)| \, P\left(\frac{r_1}{R_1}, u_1 - \theta_1\right) du_1$$

for $r_1 < R_1 < 1$. For a fixed $r_1 e^{i\theta_1}$ the left side here is subharmonic in $z_2 = r_2 e^{i\theta_2}$. Hence

$$\log^+ |f(r_1 e^{i\theta_1}, r_2 e^{i\theta_2}, \ldots, z_k)|$$

$$\leq \pi^{-1} \int_0^{2\pi} \log^+ f(r_1 e^{i\theta_1}, R_2 e^{iu_2}, \ldots, z_k) \, P\left(\frac{r_2}{R_2}, u_2 - \theta_2\right) du_2$$

for $r_2 < R_2 < 1$. This, in conjunction with the previous inequality gives

$$\log^+ |f(r_1 e^{i\theta_1}, r_2 e^{i\theta_2}, z_3, \ldots, z_k)|$$

$$\leq \pi^{-2} \int_0^{2\pi}\int_0^{2\pi} \log^+ f(R_1 e^{iu_1}, R_2 e^{iu_2}, z_3, \ldots, z_k) \, P\left(\frac{r_1}{R_1}, u_1 - \theta_1\right) P\left(\frac{r_2}{R_2}, u_2 - \theta_2\right) du_1 \, du_2$$

valid for $r_1 < R_1$, $r_2 < R_2$. Proceeding in this way we find that $\log^+|f(r_1 e^{i\theta_1}, \ldots, r_k e^{i\theta_k})|$ does not exceed the right side of (13), that is does not exceed $F_{R_1 \ldots R_k}(r_1 e^{i\theta_1}, \ldots, r_k e^{i\theta_k})$, for $r_1 < R_1, \ldots, r_k < R_k$. Making here $R_1, \ldots, R_k$ tend to 1, we find that $F(z_1, \ldots, z_k)$ majorizes $\log^+|f(z_1, \ldots, z_k)|$ in Γ_k. This completes the proof of Lemma 4.

Since $F(z_1, \ldots, z_k)$ is non-negative, it is a Poisson - Stieltjes integral (Lemma 2), and so has a restricted non-tangential limit at almost every point $(e^{i\theta_1}, \ldots, e^{i\theta_k})$ (Lemma 3). It follows that $\log^+ |f|$, and so also f, is restrictedly and non-tangentially bounded at almost every point $(e^{i\theta_1}, \ldots, e^{i\theta_k})$. Theorem 1 will therefore be automatically established if we prove Theorem 2.

Basic for the proof of Theorem 2 is the following theorem about k-harmonic functions in which, instead functions defined in a polycylinder, it is slightly easier to consider functions defined in the topological product of upper half-planes.

LEMMA 5. Let $G(w_1, \ldots, w_k)$ be a function of the complex variables $w_1 = u_1 + iv_1, \ldots, w_k = u_k + iv_k$, k-harmonic in the domain

$$v_1 > 0, \ \ldots, \ v_k > 0 \ .$$

Let us consider a set E of points $(u_1, \ldots, u_k)$ situated in the distinguished boundary

$$v_1 = 0, \ \ldots, \ v_k = 0$$

and suppose that the function G remains bounded, not necessarily uniformly, as $(w_1, \ldots, w_k)$ approaches non-tangentially any point $(u_1^o, \ldots, u_k^o) \in E$. Then G has a non-tangential limit at almost every point of E.

For the proof, see [1].

In proving Theorem 2 we shall first consider the case $k = 2$ which is fairly typical. It will be enough to add later a few explanatory remarks about the case $k > 2$.

By mapping conformally the interiors of the unit circles onto the upper half-planes we may assume that our function $f(z_1, z_2)$ is regular in the domain

$$(15) \qquad y_1 > 0, \ \ y_2 > 0 \qquad (z_1 = x_1 + iy_1, \ z_2 = x_2 + iy_2)$$

and that it is restrictedly non-tangentially bounded at every point of a set E situated in the distinguished boundary

$$(D_2) \qquad y_1 = 0, \ y_2 = 0$$

To make the situation geometrically simple, let us consider the plane X of the points (x_1, x_2) and the plane Y of the points (y_1, y_2). Every complex pair (z_1, z_2) will be treated as a pair consisting of a point (x_1, x_2) in X and of a point (y_1, y_2) in Y, and we may write the pair (z_1, z_2) in the form $(x_1, x_2; y_1, y_2)$. The set E is situated in D_2 and so its points can be written $(x_1, x_2; 0, 0)$. It will be simpler if we merely treat E as a set of points in the plane D_2. Then the situation is geometrically as follows. Let ε be a positive number less than 1, and let A_ε denote the angle in the Y plane between the rays

$$y_2 = \varepsilon y_1, \quad y_2 = \varepsilon^{-1} y_1 \ .$$

The assumption that $f(z_1, z_2)$ is restrictedly non-tangentially bounded at a point (x_1^o, x_2^o) from E implies that $f(z_1, z_2)$ remains bounded as

(16) (a) (y_1, y_2) tends to the origin $(0, 0)$ through A_ε;

(b) $|x_1 - x_1^0| \leqq A\, y_1, \quad |x_2 - x_2^0| \leqq A\, y_2$,

where A denotes an arbitrary but fixed positive constant. Condition (b) here is that of non-tangential approach. Condition (a) is that of the restrictedness of the approach and is equivalent to the condition that the ratio y_2/y_1 is contained between two positive constants. If we omit condition (a) and merely require that y_1, y_2 approach the origin through the positive quadrant in the Y plane (since $y_1 > 0$, $y_2 > 0$), we obtain the ordinary (unrestricted) non-tangential approach.

We shall now introduce two new complex variables $w_1 = u_1 + iv_1$ and $w_2 = u_2 + iv_2$. Let U and V denote respectively the planes (u_1, u_2) and (v_1, v_2). We shall establish a one-one correspondence between the planes X and U, and an analogous correspondence between the planes Y and V, by means of the formulas

$$(16) \quad \text{(a)} \quad \begin{aligned} x_1 &= u_1 + \varepsilon u_2 \\ x_2 &= \varepsilon u_1 + u_2 \end{aligned} \qquad\qquad \text{(b)} \quad \begin{aligned} y_1 &= v_1 + \varepsilon v_2 \\ y_2 &= \varepsilon v_1 + v_2 \end{aligned}$$

These four formulas can be reduced to the following two

$$(17) \qquad z_1 = w_1 + \varepsilon w_2 , \qquad z_2 = \varepsilon w_1 + w_2$$

We note that transformation (16 b) establishes one-one correspondence between the positive quadrant in the V plane and the angle A_ε in the Y plane, and that in this transformation the positive v_1 axis and the positive v_2 axis correspond to the sides of the angle A_ε. As to the correspondence between the planes X and U it is enough at this stage to observe that it is one-one. Transformation (16 a) maps the set E in the X plane into a set E^* in the U plane. The measures of the sets E need not be the same, but they differ by a multiplicative constant different from 0.

From what has just been said, follows that if we set

$$(18) \qquad f(z_1, z_2) = f(w_1 + \varepsilon w_2, \ \varepsilon w_1 + w_2) = g(w_1, w_2),$$

the function g is regular in the domain

$$(19) \qquad v_1 > 0, \quad v_2 > 0 .$$

Let (x_1^0, x_2^0) and (u_1^0, u_2^0) be two corresponding points in the planes X and U respectively. Let δ be a number greater than ε but less than 1. Thus the angle A_δ is interior to the angle A_ε. We shall need two things:

1^0 If the point (w_1, w_2) approaches non-tangentially the point (u_1^0, u_2^0), then the corresponding point (z_1, z_2) approaches (x_1^0, x_2^0) non-tangentially, and in such a way that

(20) $$\varepsilon \leqq y_2 / y_1 \leqq \varepsilon^{-1}.$$

2^0 If the point (z_1, z_2) approaches (x_1^0, x_2^0) non-tangentially and in such a way that

(21) $$\delta \leqq y_2 / y_1 \leqq \delta^{-1},$$

then the corresponding point (w_1, w_2) approaches (u_1^0, u_2^0) non-tangentially.

Owing to the homogeneous character of the mappings (16), we may assume without loss of generality that $(u_1^0, u_2^0) = (0, 0)$ and $(x_1^0, x_2^0) = (0, 0)$.

That under the assumptions of 1^0, we have condition (20) satisfied has already been pointed out, and is the main idea underlying the whole proof. If (w_1, w_2) approaches $(0, 0)$ non-tangentially, we have

$$|u_1| \leqq A\, v_1, \qquad |u_2| \leqq A\, v_2,$$

where A is a positive constant. Using this and the formulas (16 a) we get

$$|x_1| = |u_1 + \varepsilon u_2| \leqq |u_1| + \varepsilon |u_2| \leqq A\,(v_1 + \varepsilon v_2) = A\, y_1.$$

Thus $|x_1| \leqq A\, y_1$, and similarly $|x_2| \leqq A\, y_2$. This completes the proof of 1^0.

In order to prove 2^0, let us solve the equations (16). We get

$$u_1 = \frac{x_1 - \varepsilon x_2}{1 - \varepsilon^2}, \qquad v_1 = \frac{y_1 - \varepsilon y_2}{1 - \varepsilon^2}$$

and similar formulas for u_2, v_2. Remembering that $\delta \leqq y_1/y_2 \leqq \delta^{-1}$ we can write

(22) $$v_1 = \frac{y_2\left(\frac{y_1}{y_2} - \varepsilon\right)}{1 - \varepsilon^2} \geqq y_2 \frac{\delta - \varepsilon}{1 - \varepsilon^2}$$

On the other hand, assuming that $|x_1| \leqq A\, y_1$, $|x_2| \leqq A\, y_2$, we have

$$|v_1| \leq \frac{|x_1| + \varepsilon|x_2|}{1-\varepsilon^2} \leq A\,\frac{(y_1+\varepsilon y_2)}{1-\varepsilon^2}$$

$$= A\,y_2\,\frac{\left(\frac{y_1}{y_2}+\varepsilon\right)}{1-\varepsilon^2} \leq A\,y_2\,\frac{\delta^{-1}+\varepsilon}{1-\varepsilon^2}\,.$$

From this and from (22) we get that $|u_1| \leq B\,v_1$, where $B = A\,\dfrac{\delta^{-1}+\varepsilon}{\delta-\varepsilon}$. Similarly, $|u_2| \leq B\,v_2$. This completes the proof of 2°.

It is now easy to finish the proof of Theorem 2. Suppose that the function $f(z_1, z_2)$ regular in the domain (15) is restrictedly non-tangentially bounded at every point of a set E situated in the X plane. The function $g(w_1, w_2)$ defined by (18) is regular in the domain (19) and is non-tangentially (unrestrictedly) bounded at every point of the set E^* obtained from E by the mapping (16 a). By Lemma 5, the function g has a non-tangential limit almost everywhere in E^*. By proposition 2°, at almost every point of E the function $f(z_1, z_2)$ has a limit, provided z_1 tends to x_1^0 non-tangentially, z_2 tends to x_2^0 non-tangentially, and provided condition (21) is satisfied. Taking a sequence of values of δ converging to 0 we immediately obtain that the function $f(z_1, z_2)$ has a restricted non-tangential limit at almost every point of E. This completes the proof of Theorem 2 in the case $k = 2$.

Passing to the case $k > 2$, we may from the start consider a function $f(z_1, \ldots, z_k)$ regular in the domain

$$y_1 > 0, \ldots, y_k > 0. \qquad (z_j = x_j + iy_j)$$

and restrictedly and non-tangentially bounded at every point of a set E situated on the distinguished boundary

$$(D_k) \qquad y_1 = 0, \ldots, y_k = 0$$

of the domain. Let $0 < \varepsilon < 1$. We consider the transformation

$$(23) \qquad \begin{aligned} z_1 &= w_1 + \varepsilon w_2 + \varepsilon w_3 + \ldots + \varepsilon w_k \\ z_2 &= \varepsilon w_1 + w_2 + \varepsilon w_3 + \ldots + \varepsilon w_k \\ &\ldots\ldots\ldots\ldots\ldots\ldots\ldots\ldots\ldots\ldots \\ z_k &= \varepsilon w_1 + \varepsilon w_2 + \varepsilon w_3 + \ldots + w_k \end{aligned}$$

On setting $w_j = u_j + i\,v_j$ we obtain the two transformations

$$(24)\qquad \begin{aligned} x_1 &= u_1 + \varepsilon u_2 + \dots + \varepsilon u_k \\ &\cdots\cdots\cdots\cdots \\ x_k &= \varepsilon u_1 + \varepsilon u_2 + \dots + u_k \end{aligned}$$

and

$$(25)\qquad \begin{aligned} y_1 &= v_1 + \varepsilon v_2 + \dots + \varepsilon v_k \\ &\cdots\cdots\cdots\cdots \\ y_k &= \varepsilon v_1 + \varepsilon v_2 + \dots + v_k \end{aligned}$$

either of which will be denoted by T. As easily seen, T is non-singular and so it establishes a one-one correspondence between the points $(x_1, \dots, x_k)$ of the space X and the points $(u_1, \dots, u_k)$ of the space U. Let us denote by V' the "positive" octant of the space V, i. e. the set of the points $(v_1, \dots, v_k)$ with all coordinates non-negative. Then T maps V' into an "angle" A_ε, which is a conical domain with vertex at the origin 0 and situated, except for that vertex, totally in the interior of the first octant Y'. The map of A_ε by T is another "angle", which we shall denote by B_ε, and which, except for the origin, is totally in the interior of A_ε. For that $B_\varepsilon \subset A_\varepsilon$ is obvious. That the boundary points ($\pm$ 0) of B_ε, which are images of boundary points of A_ε, cannot be on the boundary of A_ε, is also clear since the boundary points of B_ε are images, through T^2, of the boundary points of the positive octant, and so cannot be T images of points from A_ε. The "angle" B_ε plays here the same role as the angle A_δ $(0 < \varepsilon < \delta < 1)$ in the proof when k was 2.

1° If the point $(w_1, \dots, w_k)$ approaches non-tangentially a point $(u_1^0, \dots, u_k^0) \in D_k$, then $(z_1, \dots, z_k)$ approaches the corresponding point $(x_1^0, \dots, x_k^0)$ restrictedly and non-tangentially.

2° If $(z_1, \dots, z_k)$ approaches $(x_1^0, \dots, x_k^0)$ non-tangentially and in such a way that $(y_1, \dots, y_k) \in B_\varepsilon$, then $(w_1, \dots, w_k)$ approaches $(u_1^0, \dots, u_k^0)$ non-tangentially.

We may assume that both $(u_1^0, \dots, u_k^0)$ and $(x_1^0, \dots, x_k^0)$ coincide with the origin.

Ad 1°. If we assume that $|u_j| \leqq A v_j$, for all j, the formulas (24) and (25) give $|x_j| \leqq A y_j$. In addition, the formulas (25) give the following inequalities valid for all points $(y_1, \dots, y_k) \in A_\varepsilon$

$$\varepsilon \operatorname*{Max}_j v_j \leqq y_h \leqq (1 + n\varepsilon) \operatorname{Max}_j v_j \qquad (h = 1, \dots, k),$$

which show that the ratio of any two y's is contained between $\varepsilon/(1 + n\varepsilon)$ and $(1 + n\varepsilon)/\varepsilon$. This proves 1°.

Ad 2°. By assumption, $|x_j| \leq A\, y_j$ for all j. Solving the equations (24) with respect to $u_1, \ldots, u_k$, we obtain u_h as a linear function of $x_1, \ldots, x_k$. Hence $|u_h| \leq B \operatorname{Max}_j |x_j|$ $(h = 1, \ldots, k)$, where B is a positive constant. This and the preceeding inequality give

$$|u_h| \leq A\, B \operatorname{Max}_j y_j \qquad (h = 1, \ldots, k)$$

and so, by (25)

$$(26) \qquad |u_h| \leq A\, B\, (1 + n\,\varepsilon) \operatorname{Max}_j v_j \, .$$

Finally, since $(y_1, \ldots, y_k) \in B_\varepsilon$, it follows that $(v_1, \ldots, v_k) \in A_\varepsilon$, and so the ratio of any two v's is $\leq (1 + \varepsilon n)\, \varepsilon^{-1}$. Applied to (26) this gives

$$|u_h| \leq A\, B\, (1 + n\,\varepsilon)^2\, \varepsilon^{-1}\, v_h \qquad (h = 1, 2, \ldots, k),$$

which completes the proof of 2°.

In the function $f(z_1, \ldots, z_k)$ let us now express the z's in terms of the w's. The resulting function $g(w_1, \ldots, w_k)$ is defined and regular for $v_1 > 0, \ldots, v_k > 0$. The restricted non-tangential boundedness of f at the points of a set E situated in the X plane implies (unrestricted) non-tangential boundedness of g at the points of a set E^* -image of E- in the U plane. By Lemma 5, g has a non-tangential limit almost everywhere in E^*, and so f has a non-tangential limit almost everywhere in E, provided $(y_1, \ldots, y_k)$ tends to the origin through B_ε. As ε approaches 0, the transformation T approaches identity, and the sets A_ε and B_ε tend to exhaust the interior of the positive octant. Taking therefore a sequence of values of ε tending to 0, we easily see that f has a restricted non-tangential limit at almost every point of E.

3. Additional remarks. (i) The proof of Theorem 2 is mainly based on Lemma 5, which is valid for k-harmonic functions, and on the transformation (23). It follows that *Theorem 2 is true for functions which are real parts of functions regular for* $y_1 > 0, \ldots, y_k > 0$. Whether Theorem 2 is valid for functions which are k-harmonic, is an open problem. A slight modification of the argument, shows that Theorem 2 remains valid for functions which are real parts of functions analytic, but not necessarily regular, for $y_1 > 0, \ldots, y_k > 0$.

(ii) Certain considerations lead to a notion which may be, roughly, described as *mixed non-tangential* convergence, and which is intermediate between restricted and unrestricted non-tangential convergence. Suppose we have a function $f(z_1, \ldots, z_k)$ defined in Γ_k, and let m be an integer not less than 2 and not exceeding k. It may happen that $f(z_1, \ldots, z_k) \to s$, as $(z_1, \ldots, z_k)$ approaches $(e^{i\theta_1^0}, \ldots, e^{i\theta_k^0})$ $(z_j = r_j e^{i\theta_j})$ non-tangentially but in such a way that the ratios

$$(27) \qquad (1 - r_j) / (1 - r_h) \qquad (1 \leq j, h \leq m)$$

remain contained between any two positive numbers, while the remaining radii r_p $(m + 1 \leq p \leq k$, if $m < k)$ tend to 1 independently of one another. This is mixed non-tangential convergence of f at the point $(e^{i\theta_1^0}, \ldots, e^{i\theta_k^0})$. In this situation we have $k - m + 1$ degrees of freedom, since one of the radii $r_1, \ldots, r_m$ is arbitrary.

Without introducing specific and precise terminology, we state the following result.

THEOREM 3. A function $f(z_1, \ldots, z_k)$ regular in Γ_k has at almost every point $(e^{i\theta_1^0}, \ldots, e^{i\theta_k^0})$ a limit of the kind just described, if

$$(28) \quad \int_0^{2\pi} \cdots \int_0^{2\pi} \log^+ |f| \, (\log^+ \log^+ |f|)^{k-m} d\theta_1 \ldots d\theta_k = O(1) \qquad (2 \leq m \leq k)$$

where $f = f(r_1 e^{i\theta_1}, \ldots, r_k e^{i\theta_k})$.

We shall only sketch the proof of the theorem. Condition (28) implies condition (12) of the preceding section, and so, by Lemma 4, the function $\log^+ |f|$ has a k-harmonic majorant in Γ_k. As the proof of Lemma 4 shows, this harmonic majorant H, which is non-negative, satisfies the condition

$$\int_0^{2\pi} \cdots \int_0^{2\pi} H \, (\log^+ H)^{k-m} \, d\theta_1 \ldots d\theta_k = O(1), \quad H = H(r_1 e^{i\theta_1}, \ldots, r_k e^{i\theta_k}).$$

It follows (see [9], Theorem 5), that $H(z_1, \ldots, z_k)$ has almost everywhere a mixed non-tangential limit of the kind described above, i. e. we assume that the ratios (27) remain bounded by two positive numbers. Hence f is bounded under similar circumstances. Theorem 3 will therefore follow if we establish the following extension of Theorem 2.

THEOREM 4. Suppose $f(z_1, \ldots, z_k)$ is regular for $y_1 > 0, \ldots, y_k > 0$ and suppose that every point of a set E situated in the space X of all points $(x_1, \ldots, x_k)$ has the following property. The function f remains bounded as $(z_1, \ldots, z_k)$ approaches any point $(x_1^0, \ldots, x_k^0) \in E$ non-tangentially and in such a way that the ratios y_j/y_h $(1 \leqq j, h \leqq m)$ remain contained between any pair of positive numbers. Then at almost every point of E, and under similar circumstances, f has a limit.

The proof is analogous to that of Theorem 2 except that, when making transformation of the type (23), we apply it only to the variables $z_1, \ldots, z_m$, and leave $z_{m+1}, \ldots, z_n$ untouched. We omit the details.

(iii) The basic transformation (23) strengthening restricted non-tangential convergence to the unrestricted one, enables us to extend to the former some properties of the latter. Without trying to achieve the greatest generality, let us consider the following situation.

THEOREM 5. *Suppose that* $f(z_1, \ldots, z_k)$ *is regular for* $y_1 > 0, \ldots, y_k > 0$, *and suppose that as* $(z_1, \ldots, z_k)$ *tends restrictedly non-tangentially to any point* $(x_1^0, \ldots, x_k^0)$ *of a set E situated on the distinguished boundary* $y_1 = 0, \ldots, y_k = 0$, *the values of f omit the interior of a certain circle* $C = C(x_1^0, \ldots, x_k^0)$ *of positive radius. Then f has a restricted non-tangential limit at almost every point of E. It is even enough to assume that the values from the interior of* $C(x_1^0, \ldots, x_k^0)$ *are omitted only in an arbitrarily small restricted and non-tangential neighborhood of* $(x_1^0, \ldots, x_k^0)$.

For the function $g(w_1, \ldots, w_k) = f(z_1 + \varepsilon z_2 + \ldots + \varepsilon z_k, \ldots, \varepsilon z_1 + \ldots + z_k)$ omits the interior of a circle $C' = C'(u_1^0, \ldots, u_k^0)$ for $(w_1, \ldots, w_k)$ approaching every point of a set E^* which is the image of E in the $(u_1, \ldots, u_k)$ space, and this is known to imply the existence of a non-tangential limit of g at almost every point of E^* (for $k = 1$, this is the very well known result of Plessner; for $k > 1$, see [1]), which, in turn, implies the existence of the restricted non-tangential limit of f at almost every point of E. A special case of the result is worth mentioning separately, namely: If the real part of a function $f(z_1, \ldots, z_k)$ regular for $y_1 > 0, \ldots, y_k > 0$, has a restricted non-tangential limit at every point of a set E situated on the distinguished boundary $y_1 = 0, \ldots, y_k = 0$, the function f has a restricted non-tangential limit at almost every point of E.

It may be added that non-tangential approach in Theorem 5 may be replaced by a slightly less stringent condition, namely that each of the points z_j approaches its limit x_j^0 through a certain fixed angle situated together with its sides in the plane $y_j > 0$, and having x_j^0 as its vertex. The angles may depend on the points x_j^0. The condition of the restrictedness of the approach is maintained. It is not difficult, however, to see that the assumption of Theorem 5 is then verified at almost every point of E. A result analogous to Theorem 5 holds for mixed non-tangential approach.

BIBLIOGRAPHY

[1] A. P. Calderón, On the behavior of the harmonic functions on the boundary, "Transactions of the American Math. Soc.," 68 (1950), pp. 47-54.

[2] O. Frostman, Potentiel d'équilibre et capacité des ensembles, "Medelanden från Lunds Universitets Matematiska Seminarium," 1935.

[3] G. H. Hardy and J. E. Littlewood, A Maximal Theorem with Function-Theoretic Applications, "Acta Math.," 54 (1930), pp. 81-110.

[4] B. Jessen, J. Marcinkiewicz and A. Zygmund, Note on the Differentiability of Multiple Integrals, "Fundamenta Mathematicae," 25 (1935), pp. 217-234.

[5] J. Marcinkiewicz and A. Zygmund, On the Summability of Double Fourier Series, "Fundamenta Mathematicae," 32 (1939), pp. 122-132.

[6] S. Saks, On the strong derivatives of functions of intervals, "Fundamenta Mathematicae," 25 (1935), pp. 235-252.

[7] A. Zygmund, "Trigonometrical Series," Warszawa, 1935.

[8] A. Zygmund, On the Differentiability of Multiple Integrals, "Fundamenta Mathematicae," 23 (1934), pp. 143-149.

[9] A. Zygmund, On the Summability of Multiple Fourier Series, "American Journal of Mathematics," 69 (1947), pp. 836-850.

[10] A. Zygmund, On the Boundary Values of Functions of Several Complex Variables, To appear in "Fundamenta Mathematicae," vol. 36.

VI. ON THE THEOREM OF HAUSDORFF-YOUNG AND ITS EXTENSIONS

By A. P. Calderón[1] and A. Zygmund

1. Introduction. Let r be any positive number. Given any interval (a, b) and a measurable function f(t) defined in it, we shall write

$$\| f \|_r = \{ \int_a^b |f|^r \, d t \}^{1/r}$$

Similarly, given any sequence $c = \{c_n\}$ of complex numbers, we shall use the notation

$$\| c \|_r = \{ \sum_n |c_n|^r \}^{1/r}$$

For any $r \geq 1$, let r' be the number given by the formula

$$r' = r / (r - 1), \quad \text{i. e.} \quad 1 / r + 1 / r' = 1.$$

By p we shall systematically denote a number satisfying the condition

$$1 < p < 2.$$

Thus $p' > 2$.

The origin of the topics discussed in this paper is the following theorem of Hausdorff-Young (see e. g. [10], Ch. IX).

THEOREM A. (i) Let $f(t) \in L^p(0, 1)$, and let

$$(1) \qquad c_n = \int_0^1 f(t) \, e^{-2\pi int} \, dt \qquad (n = 0, \pm 1, \pm 2, \dots)$$

be the Fourier coefficients of f with respect to the system

1. Fellow of the Rockefeller Foundation.

$\{e^{2\pi int}\}$. Then

(2) $$\| c \|_{p'} \leqq \| f \|_p .$$

(ii) Given any two-way infinite sequence $\{c_n\}$ of numbers such that $\| c \|_p < \infty$, there is an $f \in L^{p'}(0, 1)$ satisfying (1) and such that

(3) $$\| f \|_{p'} \leqq \| c \|_p .$$

Hardy and Littlewood, [1], completed the result by proving the following theorem.

THEOREM B. Under the assumptions of (i), we have equality in (2) if and only if $f(t) = A e^{2\pi int}$, where A is a constant and n an integer. Similarly, under the assumptions of (ii), we have equality in (3) if and only if the sequence $\{c_n\}$ has at most one element distinct from zero.

Theorem A is a special case of the following result due to F. Riesz ([2]; [10] Ch. IX) and pertaining to any system $\{\varphi_n\}$ of functions orthonormal in an interval (a , b) and satisfying there an inequality

$$|\varphi_n(t)| \leqq M \qquad \text{for all } n.$$

THEOREM C. (i) If $f \in L^p(a, b)$, then the Fourier coefficients

(4) $$c_n = \int_a^b f \bar{\varphi}_n \, dt$$

of f with respect to the system $\{\varphi_n\}$ satisfy the inequality

(5) $$\| c \|_{p'} \leqq M^{\frac{2}{p}-1} \| f \|_p$$

(ii) Given any sequence of numbers $\{c_n\}$ with $\| c \|_p$ finite, there is a function $f \in L^{p'}(a, b)$ satisfying (4) for all n and such that

(6) $$\| f \|_{p'} \leqq M^{\frac{2}{p}-1} \| c \|_p .$$

The cases of equality in (5) and (6) were investigated by Verblunsky,

[9]. To these results we revert in Section 3 below.

In the literature, there exist several proofs of Theorem C. From a certain point of view, the most interesting of them seems to be that of M. Riesz, [3]. First, because it clearly shows the origin of the theorem, and second because the idea of the proof can be applied in many other cases. The basis of the proof is a certain convexity property of the maximum of the bilinear form $\sum a_{mn} x_m y_n$. This property was later generalized by Thorin [7], [8] (see also [6]) who also showed the usefulness of the generalization.

The main purpose of this paper is to give a simple proof of the theorem of F. Riesz, and of a number of allied results, by using systematically the same tool, namely the Phragmén-Lindelöf maximum principle for analytic functions defined in a strip. That various maximum principles can be used to prove the convexity theorems of M. Riesz and Thorin, and so to contribute to the proof of F. Riesz' theorem, was a known fact (see [6], or [7]). Here, however, we use the Phragmén-Lindelöf principle directly, and as the only tool of the proof, avoiding the convexity theorems of M. Riesz and Thorin, and so simplifying the proof. The simplification is particularly noticeable in the results discussed in Sections 4 and 5 below. Earlier proofs of those results are quite laborious (see e. g. [10], Ch. IX), partly owing to the necessity of a meticulous discussion of extreme cases.

This is the plan of the present paper. Section 1 will be concluded with a few remarks concerning the Phragmén-Lindelöf principle. Section 2 contains a proof of Theorem C. Section 3 treats the cases of equality in the F. Riesz theorem. Section 4 contains a discussion of interpolation of linear operations, a topic first introduced by M. Riesz, [3]. These results contain the F. Riesz theorem as a special case, but owing to the importance and the elementary character of the latter it seemed desirable to give it a separate treatment. Section 5 is devoted to interpolation in the classes H^r. For details we refer the reader to corresponding sections.

The Phragmén-Lindelöf principle which we need here may be stated as follows.

I. Suppose that a function $f(z)$, $z = x + iy$, is continuous and bounded in a strip

$$(S) \qquad \alpha \leqq x \leqq \beta$$

and regular in the interior of S. If $|f| \leqq M$ on the border lines $x = \alpha$ and $x = \beta$, then $|f(z)| \leqq M$ also in the interior of S. If $|f(z_0)| = M$ at a point z_0 in the interior of S, then f is constant in S.

For the sake of completeness we briefly repeat the proof here. Suppose first that

(7) $$|f(x+iy)| \to 0$$

uniformly in x, $\alpha \leq x \leq \beta$, as $y \to \pm\infty$. If $z_0 = x_0 + iy_0$ is interior to S, the inequality $|f(z_0)| \leq M$ is a consequence of the Principle of Maximum applied to the rectangle $\alpha \leq x \leq \beta$, $|y| \leq \eta$, containing z_0 and with η so large that $|f(x \pm i\eta)| \leq M$ for $\alpha \leq x \leq \beta$. In the general case, we consider the function

$$f_n(z) = f(z)\, e^{z^2/n} = f(z)\, e^{(x^2-y^2)/n}\ e^{2ixy/n}$$

which satisfies condition (7) and does not exceed $M e^{\gamma^2/n}$, with $\gamma = \text{Max}(|\alpha|, |\beta|)$, on the boundary of S. Hence $|f_n(z_0)| \leq M e^{\gamma^2/n}$, and on making $n \to \infty$ we get $|f(z_0)| \leq M$.

If $|f(z_0)| = M$, then $f(z) =$ Const., since otherwise in the neighborhood of z_0 we would have points z such that $|f(z)| > M$, which is impossible.

It is often convenient to use the Phragmén-Lindelöf principle in a slightly stronger form, which is, however, an immediate consequence of (I).

II. Suppose that f(z), continuous and bounded in the strip S and regular in the interior of S, satisfies the conditions

$$|f(\alpha+iy)| \leq M_1, \qquad |f(\beta+iy)| \leq M_2 \qquad \text{for all } y.$$

Then, if L(t) is a linear function taking the values 1 and 0 for $t = \alpha$ and $t = \beta$ respectively, we have

(8) $$|f(x_0+iy_0)| \leq M_1^{L(x_0)}\ M_2^{1-L(x_0)}$$

If the sign of equality occurs here, then $f(z) = C\, M_1^{L(z)}\, M_2^{1-L(z)}$, C being a constant of modulus 1.

For it is easy to see that the function $f_1(z) = f(z)/M_1^{L(z)}\, M_2^{1-L(z)}$ satisfies the assumptions of (I), with $M = 1$.

Proposition II is often called the Three-Line Theorem.

2. Proof of the theorem of F. Riesz. We begin with case (i) of Theorem C. Let us call simple a function defined in (a, b) and taking only a finite number of values there. If (a, b) is infinite, we shall also assume that the function vanishes outside a sufficiently large interval. We may make the following three simplifying assumptions:

(a) The system $\{\varphi_n\}$ is finite;

(b) The function f is simple;

(c) $\| f \|_p = 1$.

Ad (a). If (5) is established for a finite system $\varphi_1, \varphi_2, \ldots, \varphi_N$, then making N tend to $+\infty$ we get the general result. Ad (b). The set of simple functions is dense in $L^p(a, b)$. If $f^*(t)$, with Fourier coefficients c_n^* is simple and if $\| f - f^* \|_p < \varepsilon$, then, as (4) shows, $|c_n - c_n^*|$ does not exceed

$$\| f - f^* \|_p \, \| \varphi_n \|_{p'} \leq \varepsilon \, M^{(p'-2)/p'} \left(\int_a^b |\varphi_n|^2 \, dx \right)^{1/p'} = \varepsilon M^{(p'-2)/p'}$$

Thus, if $\{\varphi_n\}$ is finite, the validity of (5) for simple functions implies its validity in the general case. Ad (c). Both sides of (5) are homogeneous of degree 1.

We now observe that

$$\| c \|_{p'} = \sum c_n d_n \tag{9}$$

where $\{d_n\}$ is a certain sequence satisfying $\| d \|_p = 1$. (As a matter of fact, we may set $d_n = |c_n|^{p'-1} \operatorname{sign} \bar{c}_n / \| c \|_{p'}^{p'-1}$). We write

$$d_n = D_n^{1/p} \, \varepsilon_n, \quad \text{with } D_n \geq 0, \quad |\varepsilon_n| = 1,$$

so that

$$\sum D_n = 1 \tag{10}$$

We also write

$$f(t) = F^{1/p}(t) \, \eta(t), \quad \text{with} \quad F(t) \geq 0, \quad |\eta(t)| = 1,$$

so that

$$\int_a^b F(t) \, dt = 1 \tag{11}$$

The coefficients c_n can now be written

$$c_n = \int_a^b F^{1/p}(t) \, \eta(t) \, \bar{\varphi}_n(t) \, dt .$$

Hence, on account of (9),

$$\| c \|_{p'} = \sum D_n^{1/p}\, \varepsilon_n \int_a^b F^{1/p}(t)\, \eta(t)\, \overline{\varphi}_n(t)\, dt .$$

We replace here $1/p$ by z and consider the function

$$\Phi(z) = \sum D_n^z\, \varepsilon_n \int_a^b F^z(t)\, \eta(t)\, \overline{\varphi}_n(t)\, dt . \tag{12}$$

Each integral on the right is a linear combination, with constant coefficients, of the exponentials of the form λ^z, where λ is positive. Hence $\Phi(z)$ is also a linear combination of such exponentials, and so is bounded in every vertical strip of finite width of the z-plane. Let us consider the upper bound of $|\Phi|$ on the lines $x = 1$ and $x = \frac{1}{2}$. In the first case,

$$|\Phi| \leq \sum D_n \int_a^b F|\varphi_n|\, dt \leq M \sum D_n \int_a^b F\, dt = M,$$

by (10) and (11). For $z = \frac{1}{2} + iy$, an application of Schwarz's inequality to the right side of (12) gives

$$|\Phi| \leq (\sum D_n)^{\frac{1}{2}} \{\sum | \int_a^b F^{\frac{1}{2}+iy}(t)\, \eta(t)\, \overline{\varphi}_n(t)\, dt|^2\}^{\frac{1}{2}} \leq 1 \quad \{\int_a^b F\, dt\}^{\frac{1}{2}} = 1,$$

by Bessel's inequality, since the integrals of the second member are Fourier coefficients of the function $F^{(1/2)+iy}(t)\, \eta(t)$.

The linear function which for $t = \frac{1}{2}$ and $t = 1$ takes respectively the values 1 and 0 is $2(1 - t)$. If therefore we apply the Phragmén-Lindelöf principle in the form II, we get

$$\| c \|_{p'} = \Phi(1/p) \leq 1^{2(1-1/p)}\, M^{(2/p)-1} = M^{(2/p)-1}$$

which completes the proof of part (i) of Theorem C.

It is well known (see e. g. [10], Ch. IX), that parts (i) and (ii) of Theorem C are easy consequences of each other. For example, in order to deduce (ii) from (i) we argue as follows. We fix an integer $N > 0$, and set $f_N = c_1 \varphi_1 + \dots + c_N \varphi_N$, where $\{c_n\}$ is the given sequence. Since φ_n belongs to $L^{p'}(a, b)$ ($\int_a^b |\varphi_n|^{p'} dt \leq M^{p'-2} \int_a^b |\varphi_n^2| dt = M^{p'-2}$), so does f_N. Therefore, for a certain g, with $\| g \|_p = 1$, we have, denoting by d_n the

Fourier coefficients of g,

$$\| f_N \|_{p'} = \int_a^b \bar{f}_N \, g \, dt = \Sigma \, \bar{c}_n d_n \leq \{ \Sigma |c_n|^p \}^{1/p} \{ \Sigma |d_n|^{p'} \}^{1/p'}.$$

On account of part (i), this gives

$$(14) \qquad \| f_N \|_{p'} \leq M^{\frac{2}{p} - 1} \left(\sum_1^N |c_n|^p \right)^{1/p}.$$

It follows that

$$\| f_{N+K} - f_N \|_{p'} \leq M^{\frac{2}{p} - 1} \left(\sum_{N+1}^{N+K} |c_n|^p \right)^{1/p} \qquad (K > 0)$$

which shows that the left side here tends to 0 as $N \to \infty$. Hence there is a function $f \in L^{p'}(a, b)$ such that $\| f - f_N \| \to 0$. Making N tend to $+\infty$, we therefore deduce from (14) the inequality (6).

It only remains to show that the c_n are the Fourier coefficients of f with respect to the φ_n. Since $p < 2$, the finiteness of $\| c \|_p$ implies that of $\| c \|_2$. The Parseval formula $\| f_{N+K} - f_N \|_2^2 = \sum_{N+1}^{N+K} |c_n|^2$ implies that f_N converges to f also in the metric L^2 (that the limit function of f_N must be the same in both metrics L^p and L^2 follows from the fact that almost everywhere $f(x) = \lim f_{N_j}(x)$, if N_j tends to $+\infty$ sufficiently rapidly). Hence, for n fixed and for $N > n$,

$$c_n = \int_a^b f_N \bar{\varphi}_n \, dt = \int_a^b f \bar{\varphi}_n \, dt + \int_a^b (f_N - f) \, \varphi_n \, dt,$$

and since the numerical value of the last term does not exceed $\| f_N - f \|_2 \cdot \| \varphi_n \|_2 = \| f_N - f \|_2$, the result follows on making N

REMARK. A direct proof of (14), from which (ii) follows, is similar to the proof of (i). Let us assume that $\sum_1^N |c_n|^p = 1$. We note that

$$(15) \qquad \| f_N \|_{p'} = \operatorname{Sup}_g \int_a^b \bar{f}_N(t) \, g(t) \, dt, \qquad (g\text{-simple}, \ \| g \|_p = 1)$$

fix one such g, and write

$$g(t) = G^{1/p}(t) \, \eta(t), \qquad \bar{c}_n = C_n^{1/p} \, \varepsilon_n,$$

where $G(t)$ and C_n are non-negative and $|\eta(t)| = 1$, $|\varepsilon_n| = 1$. Thus the integral in (15) is $\sum_1^N C_n^{1/p} \varepsilon_n \int_a^b G^{1/p} \eta \bar{\varphi}_n \, dt$, and correspondingly we introduce the function

$$\Psi(z) = \sum_1^N C_n^z \varepsilon_n \int_a^b G^z(t) \eta(t) \bar{\varphi}_n(t) \, dt,$$

which is a linear combination of exponentials λ^z, $\lambda > 0$. We again verify that on the lines $x = \frac{1}{2}$ and $x = 1$ the modulus of $|\Psi|$ does not exceed 1 and M respectively. Hence $\Psi(1/p) \leqq M^{(2/p)-1}$, and taking the Sup for all g's we get $\| f_N \|_{p'} \leqq M^{(2/p)-1}$

3. <u>The cases of equality in Theorem C</u>. If we want to investigate the cases of equality in (5) and (6), we cannot make in the foregoing proofs the simplifying assumptions (a) and (b) of Section 2. Apart from that, the proof may follow the same line, provided at each step we see when inequality can be replaced by equality. This we shall do now. The following result, due to Verblunsky, [9], generalizes Theorem B. We must assume, however, here that the system $\{\varphi_n\}$ is complete.

THEOREM D. (i) A necessary condition for equality in (5) is that

$$f(t) = \sum_{k=1}^N c_{n_k} \varphi_{n_k} \qquad (n_1 < n_2 < \dots < n_N) \tag{16}$$

For such functions we have equality in (5) if and only if

(A) $|c_{n_1}| = |c_{n_2}| = \dots = |c_{n_N}|$;

(B) $|f(t)|$ is constant in a set E of measure $1/N M^2$, and $f = 0$ outside E.

(ii) A necessary condition for the equality in (6) is that only a finite number of the c's, say $c_{n_1}, c_{n_2}, \dots, c_{n_N}$ are distinct from zero and satisfy condition A. The function f is then of the form (16), and a necessary and sufficient condition for the equality in (6) is that f satisfy condition B.

Let us begin with (i). We assume that $\| f \|_p = 1$. We know that $\| c \|_{p'}$ is finite. If we set $d_n = |c_n|^{p'-1} \operatorname{sign} \bar{c}_n / \| c \|_{p'}^{p'-1}$, we have (9). As in the proof of part (i) of Theorem C, we consider the numbers D_n, ε_n and the functions $F(t)$, $\eta(t)$. The relations (10) and (11) are satisfied. We consider the function $\Phi(z)$ defined in (12). An application of Hölder's inequality to the integrals in (12), together with (11), shows that the integrals of the moduli of the integrands are finite for z in the strip (S) $\frac{1}{2} \leqq x \leqq 1$. Moreover, as easily seen, these integrals are regular functions of z in the interior of S. Thus the terms of the series (12) are continuous in S and regular in the interior of S. An application of Hölder's inequality to the series in (12), combined with (10) and part (i) of Theorem C, shows that the series (12) converges absolutely and uniformly in S, and so $\Phi(z)$ is continuous and bounded in S, and regular in the interior of S.

If $\Phi(1/p) = \| c \|_p = M^{(2/p) - 1}$, then, by the Phragmén-Lindelöf Principle II, the function $\Phi(z)/ M^{2z-1}$ is constant in S, and in particular,

$$\Phi(1) = \Sigma D_n \varepsilon_n \int_a^b F \eta \bar{\varphi}_n \, dt = M. \tag{17}$$

Since

$$\Sigma D_n = 1, \qquad \left| \int_a^b F \eta \bar{\varphi}_n \, dt \right| \leqq M \int_a^b F \, dt = M, \tag{18}$$

it follows that <u>for any n with $D_n \neq 0$ we have</u>

$$\varepsilon_n \int_a^b F \eta \, \bar{\varphi}_n \, dt = M. \tag{19}$$

The integral on the left here is the Fourier coefficient of the integrable function $F\eta$ with respect to the function φ_n, and by the Riemann-Lebesgue theorem (valid for any uniformly bounded orthonormal system, regardless of whether (a , b) is finite or not) it tends to 0 as $n \to \infty$ (see e. g. [10], Ch. II). Hence we can find at most a finite number of D_n distinct from 0, i. e. only a finite number of c_n distinct from 0. Since $\{\varphi_n\}$ is by assumption complete, this leads to (16).

Since $\varepsilon_n = \text{sign } d_n = \text{sign } \bar{c}_n$ in $\Phi(z)$, a comparison of (17) and (19) shows that

$$\int_a^b F(t) \text{ sign } f(t) \, \bar{\varphi}_n(t) \text{ sign } \bar{c}_n \, dt = M \tag{20}$$

for $n = n_1, n_2, \ldots, n_N$. This implies two facts about the set E where F (or f) is distinct from zero. First,

$$\text{sign } f = \text{sign}(c_n \varphi_n)$$

second

$$|\varphi_n| = M,$$

both almost everywhere in E, and for $n = n_1, \ldots, n_N$. Applied to (16), this gives

$$|f(t)| = M \sum |c_{n_k}| . \tag{21}$$

Thus $|f|$ is constant almost everywhere in E (compare condition B above). Dividing both sides of the equation

$$\left| \int_E |f|^p \operatorname{sign} f \, \overline{\varphi}_n \, dt \right| = M$$

(see (19) by $(M \sum |c_{n_k}|)^{p-1}$, we get

$$\left| \int_E f \overline{\varphi}_n \, dt \right| = \left| \int_a^b f \overline{\varphi}_n \, dt \right| \leqq M^{1-p} \qquad (\sum |c_{n_k}|)^{-p},$$

for $n = n_1, \ldots, n_N$, which proves the necessity of condition A. It remains to find $|E|$. For this purpose we observe that $\int_a^b |f|^2 \, dt$, on the one hand, on account of (21), equals

$$\int_E |f|^2 \, dt = M^2 \, (\sum |c_{n_k}|)^2 \, |E| = M^2 N^2 \, |c_{n_1}|^2 \, |E|,$$

and on the other, by Parseval's formula,

$$\int_a^b |f|^2 \, dt = \sum |c_{n_k}|^2 = N \, |c_{n_1}|^2 .$$

Comparing the two members on the right, we get $|E| = 1/N M^2$.

This proves the necessity of the conditions. The verification of their sufficiency is immediate.

The proof of part (ii) of Theorem D may be obtained either by a parallel investigation of the function $\Psi(z)$ of Section 2 or by a reduction to part (i). We shall do the latter. Suppose that we have equality in (6) and that $\| c \|_p \neq 0$ (otherwise the result is obvious). Since $\| f \|_{p'}$ is finite, there is a $g(t)$ with $\| g \|_p = 1$, and with Fourier coefficients d_n such that $\| f \|_{p'} = \int_a^b \overline{f} g \, dt$. Writing $f_N = c_1 \varphi_1 + \ldots + c_N \varphi_N$, we also have $\| f - f_N \|_{p'} \to 0$.

Hence

$$(22) \quad \| f \|_{p'} = \int_a^b \bar{f}\, g\, dt = \lim \int_a^b \bar{f}_N\, g\, dt = \lim \sum_1^N \bar{c}_n d_n = \sum \bar{c}_n d_n$$

$$\leq \| c \|_p \, \| d \|_{p'} \leq M^{\frac{2}{p}-1} \, \| c \|_p \, .$$

The extreme members here are equal. Thus, in the first place, $\| d \|_{p'} = M^{(2/p)-1}$. Hence $g = d_{n_1} \varphi_{n_1} + \dots + d_{n_N} \varphi_{n_N}$, where $|d_{n_1}| = \dots = |d_{n_N}|$. Moreover $|g|$ is constant in a set E of measure $1/NM^2$, and vanishes outside E. In the second place, Hölder's inequality in (22) must degenerate into equality, which is only possible if the numbers $|c_n|^p$ and $|d_n|^{p'}$ are proportional. Thus $|c_{n_1}| = \dots = |c_{n_N}|$, and the remaining c's are zero. The first equation (22) shows that $|f|^{p'}$ and $|g|^p$ are proportional. Hence $|f| =$ Const. in E and vanishes outside. This gives the necessity of the conditions, and the sufficiency is immediate.

4. Interpolation of linear operations. Theorem C is only a special case of the general and important result of M. Riesz (see [3]) concerning the interpolation of linear operations. What is more, the proof of that general result is simple provided we introduce the notions needed to express it.

Let E be a measure space, i. e. a space in which a non-negative and totally additive measure μ is defined, at least for some ("measurable") sets. The measure $|E|$ of E may be finite or infinite. Let $r \geq 1$, and let us only consider measurable functions. We shall denote by $L^{r,\mu}$ the class of functions f defined on E and such that

$$\| f \|_{r,\mu} = \left(\int_E |f|^r \, d\mu \right)^{1/r}$$

is finite. Correspondingly, we shall write

$$\| f \|_{\infty,\mu} = \text{Ess sup } |f| \, .$$

If no confusion arises, we shall write L^r and $\| f \|_r$, simply. We shall call simple any function f taking only a finite number of values and - if $|E|$ is infinite - vanishing outside a subset of finite measure of E. The set of simple functions will be denoted by S. The set S is dense in every L^r, $1 \leq r < +\infty$, though not necessarily in L^∞ (if $|E|$ happens to be infinite).

We recall two basic facts. First of all, we have the Hölder inequality

$$\left| \int_E f\, g\, d\mu \right| \leq \| f \|_r \, \| g \|_{r'} \qquad (1 \leq r \leq +\infty).$$

Also

$$\| f \|_r = \sup_g \int_E f\, g\, d\mu \qquad \text{for } g \in S, \quad \| g \|_{r'} = 1, \quad 1 \leq r \leq +\infty .$$

Let E_1 and E_2 be two measure spaces with measures μ and ν respectively. Suppose $h = T f$ is a linear operation given for all f defined on E_1 with $\| f \|_r$ finite. The function h is supposed to be defined on E_2. We shall say that operation T is of type (r, s) where $1 \leq r \leq +\infty$, $1 \leq s \leq +\infty$, if

$$\| h \|_{s,\nu} \leq M \, \| f \|_{r,\mu} . \tag{23}$$

The least value of M here is the norm of the operation. If $T f$ is initially defined only for simple functions, and if $1 \leq r < +\infty$, then it can be extended, in a unique way, to all functions of $L^{r,\mu}$, with the preservation of the M in (23), since S is dense in L^r.

THEOREM D. Let E_1 and E_2 be two measure spaces, with measures μ and ν respectively. Let T be a linear operation defined for all simple functions f on E_1. Suppose that T is simultaneously of the types $(1/\alpha_1, 1/\beta_1)$ and $(1/\alpha_2, 1/\beta_2)$, i. e. that

$$\| T f \|_{1/\beta_1} \leq M_1 \, \| f \|_{1/\alpha_1} , \qquad \| T f \|_{1/\beta_2} \leq M_2 \, \| f \|_{1/\alpha_2}, \tag{24}$$

the points (α_1, β_1), (α_2, β_2) belonging to the square

$$0 \leq \beta \leq 1, \qquad 0 \leq \alpha \leq 1 .$$

Then T is also of type $(1/\alpha, 1/\beta)$ for all

$$\begin{aligned} \alpha &= \alpha_1(1-t) + \alpha_2 t \\ \beta &= \beta_1(1-t) + \beta_2 t \end{aligned} \qquad (0 < t < 1) \tag{25}$$

with

$$\| T f \|_{1/\beta} \leq M_1^{1-t} \, M_2^{t} \, \| f \|_{1/\alpha} . \tag{26}$$

In particular, if $\alpha \neq 0$, the operation T can be uniquely extended to the whole space $L^{1/\alpha,\mu}$, preserving (26).

Let us fix t, and so also the numbers α, β, in (25). Let us consider the functions

$$\alpha(z) = \alpha_1(1-z) + \alpha_2 z, \quad \beta(z) = \beta_1(1-z) + \beta_2 z,$$

which for $z = 0$ and $z = 1$ reduce to α_1, β_1 and to α_2, β_2 respectively. We shall consider $z = x + iy$ in the strip $0 \leq x \leq 1$. For $z = t$, $\alpha(z)$ and $\beta(z)$ reduce to α, β. For any simple f,

(27) $$\| Tf \|_{1/\beta} = \operatorname{Sup}_{g} \int_{E_2} Tf \cdot g \, d\nu \qquad (g\text{-simple}, \ \| g \|_{1/1-\beta} = 1).$$

We may assume that $\| f \|_{1/\alpha} = 1$. We fix f and g,

$$f = |f| \, e^{iu}, \qquad g = |g| \, e^{iv}$$

and consider the integral

(28) $$I = \int_{E_2} Tf \cdot g \, d\nu.$$

Assuming temporarily that $\alpha > 0$, $\beta < 1$, we introduce the simple functions

(29) $$F_z = |f|^{\frac{\alpha(z)}{\alpha}} e^{iu}, \qquad G_z = |g|^{\frac{1-\beta(z)}{1-\beta}} e^{iv},$$

depending on the parameter z, and the integral

(30) $$\Phi(z) = \int_{E_2} T F_z \cdot G_z \, d\nu$$

which reduces to I for $z = t$.

In (29) we assume that $f^{\alpha(z)/\alpha} = 0$ wherever $f = 0$, regardless of the value of the exponent. Similarly for G_z. Thus if $c_1, c_2, \ldots$ are the distinct from zero values of f, and $\chi_1, \chi_2, \ldots$ the characteristic functions of the sets where these values are taken, and if the similar quantities for g are denoted by $c_1', c_2', \ldots, \chi_1', \chi_2', \ldots$, we have (with $c_j = |c_j| \, e^{iu_j}$, $c_k' = |c_k'| \, e^{iv_k}$)

$$F_z = \sum e^{iu_j} |c_j|^{\frac{\alpha(z)}{\alpha}} \chi_j, \quad G_z = \sum e^{iv_k} |c_k'|^{\frac{1-\beta(z)}{1-\beta}} \chi_k'$$

$$T F_z = \sum |c_j|^{\frac{\alpha(z)}{\alpha}} e^{iu_j} T \chi_j$$

Hence

$$\Phi(\ ) = \sum |c_j|^{\alpha(z)/\alpha} |c'_k|^{\frac{1-\beta(z)}{1-\beta}} e^{i(u_j+v_k)} \int_{E_2} T \chi_j \cdot \chi_k' \, d\nu$$

is a linear combination of exponentials λ^z with $\lambda > 0$.

Let us consider any z whose real part is zero. Thus the real parts of $\alpha(z)$, $\beta(z)$ are α_1, β_1. Hölder's inequality applied to (30) gives

$$|\Phi(z)| \leqq \| T F_z \|_{1/\beta_1} \| G_z \|_{1/(1-\beta_1)}$$

(31)

$$\leqq M_1 \| F_z \|_{1/\alpha_1} \| G_z \|_{1/(1-\beta_1)}$$

On account of (29) one has

$$\| F_z \|_{1/\alpha_1} = \| |f|^{\frac{\alpha_1}{\alpha}} \|_{1/\alpha_1} = \| f \|^{\frac{\alpha_1}{\alpha}}_{1/\alpha} = 1^{\frac{\alpha_1}{\alpha}} = 1,$$

the second equation here being valid both for $\alpha_1 > 0$ and $\alpha_1 = 0$. Similarly

$$\| G_z \|_{1/(1-\beta_1)} = \| |g|^{\frac{1-\beta_1}{1-\beta}} \|_{1/(1-\beta_1)} = \| g \|^{\frac{1-\beta_1}{1-\beta}}_{1/(1-\beta)} = 1.$$

Hence, by (31), we have $|\Phi(z)| \leqq M_1$ on the line $x = 0$. Similarly, $|\Phi(z)| \leqq M_2$ on the line $x = 1$. Hence $I = \Phi(t) \leqq M_1^{1-t} M_2^t$. On account of (27), we get (26).

It still remains to consider the cases $\alpha = 0$ and $\beta = 1$ (which are without interest in applications). If e.g. $\beta = 1$, then also $\beta_1 = \beta_2 = 1$, and $\alpha > 0$. We define F_z as before, and set $G_z = g$ (so that G_z is independent of z). The rest of the proof is simplified by the fact that in (31) we may replace $\| G_z \|_{1/(1-\beta_1)}$ by $\| g \|_{1/(1-\beta)} = 1$. If $\alpha = 0$, and so $\alpha_1 = \alpha_2 = 0$, $\beta < 1$, we define F_z as f and keep the old definition of G_z

(one can also easily see that in this second exceptional case, and more generally in every case when $\alpha_1 = \alpha_2$, the inequality (26) follows from (24) by a simple application of Hölder's inequality).

This completes the proof of Theorem E.

REMARKS. 1° In the definition of the norm $\| f \|_{r,\mu}$ we may assume that $r > 0$, though for $0 < r < 1$ the norm so defined does not satisfy the triangle inequality. The set S is dense in $L^{r,\mu}$ even for $0 < r < 1$. Since in the foregoing proof Hölder's inequality is applied only in connection with the exponents $1/\beta$, and not with $1/\alpha$, it is immediately seen that Theorem E remains valid for the strip $0 \leq \alpha < +\infty$, $0 \leq \beta \leq 1$. The extension, however, to the case $\alpha > 1$ seems to have no interesting application. The situation here is different from that in the case of power series, where the classes $H_{1/\alpha}$ are of interest even if $\alpha > 1$. See Section 5 below.

2° Theorem E can be extended to multilinear operations, i. e. to operations

$$T\,[f_1, f_2, \ldots, f_n]$$

linear in each variable f_j. Theorem F that follows will be needed in the next section, and for this reason we give here a precise statement and a sketch of proof which closely follows that of Theorem E.

THEOREM F. Let E and E_1, E_2, ..., E_n be measure spaces with measures ν and μ_1, μ_2, ..., μ_n respectively. Let $h = T\,[f_1, f_2, \ldots, f_n]$ be a multilinear operation defined for simple functions f_j on E_j, $j = 1, 2, \ldots, n$. The functions h are defined on E. Suppose that T is simultaneously of the types $(1/\alpha_1^{(1)}, \ldots, 1/\alpha_n^{(1)}, 1/\beta^{(1)})$ and $(1/\alpha_1^{(2)}, \ldots, 1/\alpha_n^{(2)}, 1/\beta^{(2)})$, that is that

$$(32) \qquad \| T[f_1, f_2, \ldots, f_n] \|_{1/\beta^{(k)}} \leq M_k \, \| f_1 \|_{1/\alpha_1^{(k)}} \cdots \| f_n \|_{1/\alpha_n^{(k)}} \qquad (k = 1, 2)$$

where

$$0 \leq \beta^{(k)} \leq 1, \qquad 0 \leq \alpha_j^{(k)} \leq 1 \qquad (k = 1,2;\ j = 1,2, \ldots, n)$$

Then T is also of the type $(1/\alpha_1, \ldots, 1/\alpha_n, 1/\beta)$ for

$$(33) \qquad \beta = (1-t)\beta^{(1)} + t\beta^{(2)}, \quad \alpha_j = (1-t)\alpha_j^{(1)} + t\alpha_j^{(2)} \qquad (0 < t < 1),$$

and the inequality

$$(34) \qquad \| T[f_1, f_2, \ldots, f_n] \|_{1/\beta} \leq M_1^{1-t} M_2^t \| f_1 \|_{1/\alpha_1} \cdots \| f_n \|_{1/\alpha_n}$$

holds. In addition, if all the α_j are positive, T can be extended by continuity to $L_{1/\alpha_1} \times L_{1/\alpha_2} \times \ldots \times L_{1/\alpha_n}$, preserving (34).

Let us first suppose that $\alpha_1, \alpha_2, \ldots, \alpha_n$ are positive and that $\beta < 1$. Let us fix simple functions $f_1, f_2, \ldots, f_n$ with $\| f_j \|_{1/\alpha_j, \mu_j} = 1$, $j = 1, 2, \ldots, n$, and a simple function g with $\| g \|_{1/1-\beta, \nu} = 1$. We fix t in (33) and consider the functions

$$\beta(z) = \beta^{(1)}(1-z) + \beta^{(2)} z$$

$$\alpha_j(z) = \alpha_j^{(1)}(1-z) + \alpha_j^{(2)} z ,$$

reducing to $\beta, \alpha_1, \ldots, \alpha_n$ for $z = t$. Writing $f_j = |f_j| e^{iu_j}$, $g = |g| e^{iv}$ we consider the integral

$$(35) \quad \Phi(z) = \int_E T[|f_1|^{\frac{\alpha_1(z)}{\alpha_1}} e^{iu_1}, \ldots, |f_n|^{\frac{\alpha_n(z)}{\alpha_n}} e^{iu_n}] \, |g|^{\frac{1-\beta(z)}{1-\beta}} e^{iv} \, d\nu,$$

which for $z = t$ reduces to

$$I = \int_E T[f_1, \ldots, f_n] \, g \, d\nu .$$

Since g and the f_j are simple functions, $\Phi(z)$ is a linear combination, with constant coefficients, of exponential functions λ^z, $\lambda > 0$. For $x = 0$, Holder's inequality and (32) (for $k = 1$) give

$$|\Phi(z)| \leq \| |g|^{\frac{1-\beta^{(1)}}{1-\beta}} \|_{1/(1-\beta^{(1)})} \cdot \| T[\ldots, |f_j|^{\frac{\alpha_j(z)}{\alpha_j}} e^{iu_j}, \ldots] \|_{1/\beta^{(1)}}$$

$$\leq 1 \cdot M_1 \prod_j \| |f_j|^{\frac{\alpha_j^{(1)}}{\alpha_j}} \|_{1/\alpha_j^{(1)}} = M_1 .$$

Similarly, $|\Phi(z)| \leq M_2$ for $x = 1$. Hence $I = \Phi(t) \leq M_1^{1-t} M_2^t$. Since the upper bound of I for all simple g's with $\| g \|_{1/(1-\beta)} = 1$ gives $\| T[f_1, \ldots, f_n] \|_{1/\beta}$, (34) follows when $\| f_j \|_{1/\alpha_j} = 1$ for all j, and so also for all simple f_j.

The exceptional case $\beta = 1$ is treated as before by replacing the expression $|g|^{\frac{1-\beta(z)}{1-\beta}} e^{iv}$ in (35) by g. Similarly, if some of the α_j are zero, we replace the corresponding $|f_j|^{\alpha_j(z)/\alpha_j} e^{iu_j}$ in (35) by f_j.

It remains to show that if all the α_j are positive, and if (34) is valid for simple f_j, then T can be extended by continuity to $L_{1/\alpha_1} X \ldots X L_{1/\alpha_n}$. This will follow from the inequality

$$\| T [f_1^{(1)}, \ldots, f_n^{(1)}] - T [f_1^{(2)}, \ldots, f_n^{(2)}] \|_{1/\beta}$$

$$\leq M_1^{1-t} M_2^t \left(\sum_j \| f_j^{(1)} - f_j^{(2)} \|_{1/\alpha_j} \right) \left(\operatorname{Sup}_{j,k} \| f_j^{(k)} \|_{1/\alpha_j} \right)^{n-1}$$

and the latter is, in turn, a consequence of the inequality

$$\| T [f_1^{(1)}, f_2^{(1)}, \ldots, f_n^{(1)}] - T [f_1^{(2)}, f_2^{(2)}, \ldots, f_n^{(2)}] \|_{1/\beta}$$

$$\leq \| T [f_1^{(1)}, f_2^{(1)}, \ldots, f_n^{(1)}] - T [f_1^{(2)}, f_1^{(1)}, \ldots, f_n^{(1)}] \|_{1/\beta}$$

$$(35a) \qquad + \| T [f_1^{(2)}, f_1^{(1)}, \ldots, f_n^{(1)}] - T [f_1^{(2)}, f_2^{(2)}, \ldots, f_n^{(1)}] \|_{1/\beta}$$

$$+ \cdots\cdots\cdots\cdots\cdots\cdots\cdots$$

$$+ \| T [f_1^{(2)}, \ldots, f_{n-1}^{(2)}, f_n^{(1)}] - T [f_1^{(2)}, f_2^{(2)}, \ldots, f_n^{(2)}] \|_{1/\beta}$$

5. Interpolation of linear operations in the classes H^r. We begin by recapitulating familiar facts from the theory of the classes H^r. The details can be found e. g. in [10] Chapter VII.

Let

$$(35b) \qquad F(z) = \sum_0^\infty c_n z^n$$

be a function regular inside the unit circle $|z| < 1$. For any $r > 0$ we write

$$M_r(\rho) = M_r(\rho ; F) = \left\{ \frac{1}{2\pi} \int_0^{2\pi} |F(\rho e^{i\theta})|^r d\theta \right\}^{1/r} \qquad (0 \leq \rho < 1)$$

Here $M_r(\rho)$ is a non-decreasing function of ρ, so that the limit,

(37) $$\| F \|_r = \lim_{\rho \to 1} M_r(\rho; F)$$

exists. By H^r we shall mean the class of functions $F(z)$ for which $\| F \|_r$ is finite. If F belongs to H^r, the nontangential limit $F(e^{i\theta}) = \lim_{z \to e^{i\theta}} F(z)$ exists for almost every θ, and is of the class L^r over the interval $(0, 2\pi)$. Moreover, as $\rho \to 1$, $M_r(\rho; F)$ tends to $\{(2\pi)^{-1} \int_0^{2\pi} |F(e^{i\theta})|^r d\theta\}^{1/r}$, so that, except for an irrelevent numerical factor $(2\pi)^{-1/r}$, the norm $\| F \|_r$ of $F(z)$ coincides with the previously introduced, in Section 1, norm $\| F \|_r$ of $F(e^{i\theta})$.

Also

$$\left\{\frac{1}{2\pi} \int_0^{2\pi} |F(e^{i\theta}) - F(R e^{i\theta})|^r d\theta \right\}^{1/r} \longrightarrow 0$$

as $R \to 1$. Therefore, if we fix an R sufficiently close to 1, and then an integer N sufficiently large, the polynomial $p(z) = \sum_0^N c_n R^n z^n$ will satisfy the inequality

$$\left\{\frac{1}{2\pi} \int_0^{2\pi} |F(e^{i\theta}) - p(e^{i\theta})|^r d\theta \right\}^{1/r} \leqq \varepsilon$$

and so also the inequality

$$\| F - p \|_r = \lim_{\rho \to 1} \left\{\frac{1}{2\pi} \int_0^{2\pi} |F(\rho e^{i\theta}) - p(\rho e^{i\theta})|^r d\theta \right\}^{1/r} \leqq \varepsilon$$

Thus, with metric (37), the set of all polynomials

$$p(z) = d_0 + d_1 z + \dots + d_n z^n$$

is dense in H^r. Since the $p(z)$ belong to H^r, class H^r may be defined as the closure, under metric (37), of all polynomials $p(z)$.

For $0 < r < 1$, metric (37) does not satisfy the triangle inequality, and so H^r is not a Banach space, though it is a complete metric space, if the distance $d(F, G)$ between $F(z)$ and $G(z)$ is defined as $\| F - G \|_r^r$. For our purposes, however, a norm need not satisfy the triangle inequality.

Let $B(z)$ be the Blaschke product corresponding to an $F(z) \in H^r$. Then $F(z) = B(z)\, G(z)$, where $G(z)$ has no zeros for $|z| < 1$. Since $|B(e^{i\theta})| = 1$ almost everywhere, we have $\| F \|_r = \| G \|_r$. It is sometimes convenient to have G(0) real and positive, and to use the decomposition

(37a) $$F(z) = e^{i\gamma} B(z) G(z)$$

where γ and B(0) are real, and G(0) positive.

If $r > 1$, a function F(z) belongs to H^r if and only if the real part of F(z) is the Poisson integral of a real-valued $f \in L^r(0, 2\pi)$. If, in addition, Im F(0) = 0, we obtain the formula

(38) $$F(z) = \frac{1}{2\pi} \int_0^{2\pi} \frac{e^{it} + z}{e^{it} - z} f(t)\, dt$$

Then

(39) $$\| F \|_r \leq A_r \| f \|_r,$$

where A_r is a constant depending on r only (M. Riesz' inequality). It immediately follows that (39) remains valid (with A_r replaced by $2 A_r$, which is irrelevant for us) if f in (38) is complex-valued.

Let us consider any linear operation h = TF defined for all F in H^r, yielding $h \in L^s$ (in some measure space E) and satisfying

(40) $$\| h \|_s \leq M \| F \|_r,$$

with M independent of F. If T is initially defined in a linear subset of functions F dense in H^r and satisfying (40), we can extend operation T to the whole of H^r, preserving the constant M. The polynomials are an example of a linear subset dense in every H^r. Our problem is to obtain for the classes H^r an interpolation theorem analogous to Theorem E. For $r > 1$, there is little difficulty here on account of the correspondence established by (38) between the classes H^r and L^r. But this correspondence breaks down for r=1, and it so happens that in this case the power series possess interesting properties which we might like to interpolate and which are not shared by functions which are merely integrable. That interpolation in such cases is still possible was first shown, on a particular example, by Thorin [7]. Another example, and a more general treatment will be found in Salem and Zygmund [5]. Here we are not interested in particular examples of interpolation but want to obtain, so far as it is possible, an extension of Theorem E to classes H^r.

THEOREM G. Let (α_1, β_1) and (α_2, β_2) be two points of the strip

$$0 < \alpha < +\infty, \qquad 0 \leq \beta \leq 1.$$

Let T be a linear operation defined for all polynomials p, whose values are measurable functions in a measurable space E, and such that

$$(41)\qquad \begin{aligned} \| T p \|_{1/\beta_1} &\leqq M_1 \| p \|_{1/\alpha_1} \\ \| T p \|_{1/\beta_2} &\leqq M_2 \| p \|_{1/\alpha_2} \end{aligned}$$

Then for every point (α, β) of the segment

$$(42)\qquad \alpha = \alpha_1(1-t) + \alpha_2 t, \qquad \beta = \beta_1(1-t) + \beta_2 t \qquad (0 < t < 1)$$

we have the inequality

$$(43)\qquad \| T p \|_{1/\beta} \leqq K M_1^{1-t} M_2^{t} \| p \|_{1/\alpha},$$

K denoting a constant depending on α_1, α_2 only. In particular, T can be extended to the whole space $H_{1/\alpha}$ preserving (43).

The inequalities (41) define TF both in H_{1/α_1} and H_{1/α_2}, with the same M_1 and M_2 respectively. Moreover, if F belongs simultaneously to H_{1/α_1} and H_{1/α_2}, the value of TF obtained by the extension of the operation is the same in both cases. For suppose that

$$(44)\qquad \alpha_1 \leqq \alpha_2 .$$

Then $H_{1/\alpha_1} \subset H_{1/\alpha_2}$. If $F \in H_{1/\alpha_1}$ and $\| p_n - F \|_{1/\alpha_1} \to 0$, then also $\| p_n - F \|_{1/\alpha_2} \to 0$ (on account of (44), we have $\| G \|_{1/\alpha_2} \leqq \| G \|_{1/\alpha_1}$ for any G(z)). The inequalities (41), extended to H_{1/α_1} and H_{1/α_2} show that $T p_n$ tends to limits both in L_{1/α_1} and L_{1/α_2}. These limits must be the same because they are, almost everywhere, ordinary limits of a suitable subsequence of the sequence $\{T p_n\}$.

Let now n be a positive integer such that $\alpha_2/n < 1$ (compare (44)). Let us consider any simple complex-valued functions $g_1, g_2, \ldots, g_n$ in the interval $(0, 2\pi)$ and let us set

$$(45)\qquad F_j(z) = \frac{1}{2\pi} \int_0^{2\pi} \frac{e^{it} + z}{e^{it} - z} g_j(t)\, dt \qquad (j = 1, \ldots, n).$$

Let us write

(46) $$T^*[g_1, g_2, \dots, g_n] = T[F_1 \cdot F_2 \cdot \dots \cdot F_n]$$

The functions $F_1, F_2, \dots, F_n$ belong to every H^r, and so does their product. In particular, the latter belongs to both classes H_{1/α_1} and H_{1/α_2}, for which T is already defined. Thus T^* is a multilinear operation defined for all simple functions $g_1, g_2, \dots, g_n$.

By Hölder's inequality

(47) $$\| F_1 \cdots F_n \|_{1/\alpha_k} \leq \| F_1 \|_{n/\alpha_k} \cdots \| F_n \|_{n/\alpha_k} \qquad (k = 1, 2).$$

Denoting by A_r the constant in (39), for f complex-valued, we have

(48) $$\| F_j \|_{n/\alpha_k} \leq A_{n/\alpha_k} \| g_j \|_{n/\alpha_k} \qquad (j = 1, \dots, n;\ k = 1, 2).$$

Hence, starting with (46), then using (41) in the whole of H_{1/α_1} and H_{1/α_2} respectively, and finally applying (47) and (48), we get

(49) $$\| T^*[g_1, \dots, g_n] \|_{1/\beta_k} \leq M_k A^n_{n/\alpha_k} \| g_1 \|_{n/\alpha_k} \cdots \| g_n \|_{n/\alpha_k}$$

for $k = 1, 2$. An application of Theorem E gives

(50) $$\| T^*[g_1, \dots, g_n] \|_{1/\beta} \leq | A^{1-t}_{n/\alpha_1} A^t_{n/\alpha_2} |^n M_1^{1-t} M_2^t \prod_j \| g_j \|_{n/\alpha}.$$

Formula (46) defines T^* when $g_1, \dots, g_n$ are simple. The formulas (49) show that T^* can be extended to $L^{n/\alpha_k} \times \dots \times L^{n/\alpha_k}$ $(k = 1, 2)$ and that the extension satisfies (49). But if $g_j \in L^{n/\alpha_k}$, then the F_j in (45) belongs to H^{n/α_k}. Hence, by (47), $F_1 \dots F_n$ belongs to H^{1/α_k}, which means that $T[F_1 \dots F_n]$ is defined. We must show that (46) holds for the extended T^*. For if the g_j belong to L^{n/α_k}, and if g_j^m are simple functions such that $\| g_j^m - g_j \|_{n/\alpha_k} \to 0$ as $m \to \infty$ $(k = 1, 2;\ j = 1, \dots, n)$ then

(51) $$\| T^*[g_1^m, \dots, g_n^m] - T^*[g_1, \dots, g_n] \|_{1/\beta_k} \to 0$$

by an inequality analogous to (35a). On the other hand, by (39), if F_j^m is derived from g_j^m by means of (45), we have $\| F_j^m - F_j \|_{n/\alpha_k} \to 0$,

$\| F_j^m \|_{n/\alpha_k} \leq A_{n/\alpha_k} \| g_j^m \| = o(1)$, so that, as in (51), (but using (41) in the proof)

$$\| T [F_1^m \dots F_n^m] - T [F_1 \dots F_n] \|_{1/\beta_k} \longrightarrow 0,$$

which proves (46) in the cases considered.

We are now going to prove that for a fixed polynomial p we have (43). Let us apply to p the decomposition (37a). Without losing generality we may assume that $\gamma = 0$. The function G is now a polynomial. Thus, if we write

$$(52) \qquad p = F_1 \dots F_n, \quad \text{where } F_1 = B\, G^{1/n}, \quad F_2 = \dots = F_n = G^{1/n}$$

the functions F_j are bounded, and so also of the class H^{n/α_1}. Assuming that $\operatorname{Im} G^{1/n}(0) = 0$, we may represent the F_j by the formulas (45); where the g_j are real-valued and of the class L^{n/α_1}. Hence

$$(53) \qquad T\, p = T\, [F_1 \dots F_n] = T^* [g_1, \dots, g_n]$$

The functions g_j also belong to $L^{n/\alpha}$ (because $\alpha \geq \alpha_1$, or simply because they belong to every L^r, $r > 0$). But the formula (50), which was initially established for g_j simple shows that the operation T can be extended to $L^{n/\alpha} \times L^{n/\alpha} \times \dots \times L^{n/\alpha}$, with the preservation of (50). Combining (53) with (50) we get

$$\| T\, p \|_{1/\beta} = \| T^* [g_1, \dots, g_n] \|_{1/\beta}$$

$$\leq |A_{n/\alpha_1}^{1-t} \quad A_{n/\alpha_2}^{t}|^n \; M_1^{1-t} M_2^t \; \prod_j \; \left| \int_0^{2\pi} |g_j(t)|^{n/\alpha} \, dt \right|^{\alpha/n}$$

The last product $\prod$ here does not exceed

$$\prod_j \left| \int_0^{2\pi} |F_j(e^{it})|^{n/\alpha} \, dt \right|^{\alpha/n} = \prod_j \left| \int_0^{2\pi} |G\,(e^{it})|^{n/\alpha} \, dt \right|^{\alpha/n} = (2\pi)^{1/\alpha} \|p\|_{1/\alpha}$$

see (52), which gives (43) with $K = (2\pi)^{1/\alpha_1} B^n$, where $B = \operatorname{Max}(A_{n/\alpha_1}, A_{n/\alpha_2})$.

REMARKS. 1° It is important to stress the fact that in the definition of the operation $T^*[g_1, \dots, g_n]$ given by (46) it was necessary to consider complex-valued g_j, though in the decomposition (52) we only have real-valued g_j. The point is that in the proof of Theorem F it is essential to assume that the simple functions used are complex-valued.

2° In Theorem G for the functions p for which the operation T is initially defined we may take, instead of polynomials, any linear system of functions which is dense both in H_{1/α_1} and H_{1/α_2}. This is, however, no generalization of the theorem since, on account of (41), T is immediately extensible to all polynomials both in H_{1/α_1} and H_{1/α_2}.

3° The presence of the constant K, about whose best numerical value we know nothing, makes Theorem G slightly different from Theorems E and F.

4° The applications of the main result of (5), can be obtained, with even greater ease, from Theorem G.

Bibliography

[1] G. H. Hardy and J. E. Littlewood, Some new properties of Fourier constants, "Math. Annalen," 97 (1926), pp. 159-209.

[2] F. Riesz, Ueber eine Verallgemeinerung des Parsevalschen Formel, "Math. Zeitschrift," 18 (1923), pp. 117-124.

[3] M. Riesz, Sur les maxima des formes billinéaires et sur les fonctionelles linéaires, "Acta Mathematica," 49 (1926), pp. 465-497.

[4] R. Salem, Convexity theorems, "Bull. of the American Math. Society," 55 (1949), pp. 851-860.

[5] R. Salem and A. Zygmund, A convexity theorem, "Proc. of the National Academy of Sciences," Vol. 34, No. 9, (1948), pp. 443-447.

[6] J. D. Tamarkin and A. Zygmund, Proof of a theorem of Thorin, "Bull. of the American Math. Society," 50 (1944), pp. 279-282.

[7] G. O. Thorin, "Convexity theorems," Uppsala 1948, pp. 1-57.

[8] G. O. Thorin, An extension of a convexity theorem due to M. Riesz, "Kungl. Fysiografiska Saellskapets i Lund Forhaendliger," 8 (1939), nr. 14.

[9] S. Verblunsky, Fourier constants and Lebesgue classes, "Proc. London Math. Soc.," (2) 39, pp. 1-31.

[10] A. Zygmund, "Trigonometrical series," Warsaw, 1935.

GPSR Authorized Representative: Easy Access System Europe - Mustamäe tee 50, 10621 Tallinn, Estonia, gpsr.requests@easproject.com

www.ingramcontent.com/pod-product-compliance
Lightning Source LLC
Chambersburg PA
CBHW021048210326
41598CB00016B/1132
* 9 7 8 0 6 9 1 0 7 9 3 0 1 *